Textile Science and Clothing Technology

Series editor

Subramanian Senthilkannan Muthu, Kowloon, Hong Kong

More information about this series at http://www.springer.com/series/13111

Subramanian Senthilkannan Muthu
Editor

Sustainable Fashion: Consumer Awareness and Education

 Springer

Editor
Subramanian Senthilkannan Muthu
Head of Sustainability
SgT group and API
Kowloon, Hong Kong

ISSN 2197-9863   ISSN 2197-9871   (electronic)
Textile Science and Clothing Technology
ISBN 978-981-13-4596-8   ISBN 978-981-13-1262-5   (eBook)
https://doi.org/10.1007/978-981-13-1262-5

Printed on acid-free paper

This Springer imprint is published by the registered company Springer Nature Singapore Pte Ltd.
part of Springer Nature
The registered company address is: 152 Beach Road, #21-01/04 Gateway East, Singapore 189721,
Singapore

This book is dedicated to:
The lotus feet of my beloved
Lord Pazhaniandavar
My beloved late Father
My beloved Mother
My beloved Wife Karpagam and
Daughters—Anu and Karthika
My beloved Brother
Last but not least
To everyone working in the fashion
sector to make it SUSTAINABLE

Contents

Consumer's Awareness on Sustainable Fashion

R. Rathinamoorthy

Abstract The impact of textile and fashion products on the environment is huge and these industries are known as the second most polluting industry in the world, next to oil industries. However, the knowledge about the impact of textile manufacturing methods is still not known to the end users or customers. It also important to note that around 1% of the clothing materials produced only recycled completely towards sustainable production. Researchers also mentioned that, previous studies estimated that more than half of fast fashion items produced was disposed of in under a year. As the momentum towards the sustainable fashion increased in recent years, the consumer's knowledge on product is the key for the technologies to sustain. Hence, in this research work, a survey conducted among the individuals in the age group of 20–35, mostly college students and young entrepreneurs and employees from Tamil Nadu, India, to analyse their knowledge on the sustainability concept. From the results of the analysis it is noted that the customers are aware of the environmental implication of apparel manufacturing. However, the customers do not have the moral attitude to engage in a sustainable and ethical purchase. The customers attitude, lifestyle are the major influencing factor over knowledge. In consolidate the sustainable knowledge level observed high but the purchase behavior of the customer not improved significantly as expected due to the external influencing factors.

Keywords Fast fashion · Sustainable fashion · Customer knowledge
Preferences · Internal and external factors

1 Introduction

Fashion industry consists of different market sectors. All the different sectors of the industry can be grouped under two category called "haute couture" and "ready-made" sector. The first one is high end fashion materials manufactured for custom fitted

R. Rathinamoorthy (✉)
Department of Fashion Technology, PSG College of Technology, Coimbatore, India
e-mail: r.rathinamoorthy@gmail.com

© Springer Nature Singapore Pte Ltd. 2019
S. S. Muthu (ed.), *Sustainable Fashion: Consumer Awareness and Education* ,
Textile Science and Clothing Technology,
https://doi.org/10.1007/978-981-13-1262-5_1

measurements and the second on is focused on standardized clothing sizes (Hines and Bruce 2007). Other then these two mentioned category, a section which falls in the intermediate region of high fashion section and ready to wear category is called as Fast fashion. Fast fashion is a term used to describe clothing collections that are based on the most recent fashion trends. These are generally adaption from current high fashion luxury trend. The main advantage of the fast fashion system is its fast turnaround time that encourages disposability. This is a concept dominated by consumption, fast-changing trends, and low quality; leading consumers buy more clothes because they are affordable but discard these after only one season (Rathinamoorthy 2018).

The fast fashion is one of the major driving forces for the young customer's selection of apparels. It influencing the customers culture and buying behavior and makes them to purchase more amount of the apparels regardless of their needs. Another one of the important concept which is frequently talked in the industry is "sustainability". Fast fashion and sustainability are two contradictory concepts often mentioned in the fashion world together. The fast fashion concept in the market had resulted in excessive textile consumption and the disposal of it before their life time (Cao et al. 2014).

As a result the fast fashion concept motivates the manufactures to produce the clothing in a mass quantity and encourages the over consumption. Ultimately the fast fashion reaches the market with very less turnaround time; hence, it also motivates the customers to purchase new clothing frequently by disposing the available one at the earliest. This model provides huge opportunities for profit and innovation in the fashion manufacturers. Zara, H&M and Forever 21 are examples of fast fashion stores. These brands are mainly powered by the Internet, globalization and technological innovations (Pal 2016). These stores offers new product after every month or every two weeks instead of old season timing like twice in a year. This fast fashion trend became perfect choice for the shoppers of different income level due to its constant turnover and affordable price range (Siegle 2011) (Fig. 1).

2 Sustainability

The concept of sustainability gained momentum from the Brundtland Report and the Rio Summit. The concept made a significant impact in the society. The sustainable concept has been adapted as a policy goal by many institutions, governments, businesses, and civil society. The fundamental importance for the widespread adoption of the concept is its flexible nature. As a concept, its malleability allows it to be rather open, dynamic, and evolving, and applicable to a variety of situations across space and time. Its openness allows participants at multiple levels and across sectors and institutions to reinterpret it for their particular situation (Kates et al. 2018).

In General the sustainable development or sustainability can be defined as,

> A development that meets present needs without compromising the ability of future generations to meet their own needs.

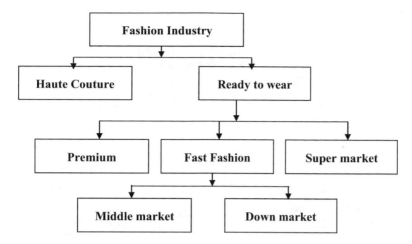

Fig. 1 Market segmentation of the fashion industry

The concept of sustainable development is built on three pillars, which represent three fundamental objectives (Fig. 2),[1]

1. Continue to produce riches in order to satisfy the needs of the world's population (economic pillar)
2. Aim to reduce inequalities between the peoples of the world (social pillar)
3. Avoid degrading the environment that future generations will inherit (environmental pillar).

(i) Social

Sustainability in societal aspect called as social sustainability. It emphasis the acts like, peace, social justice, human rights, children's rights, labor rights, gender equality, reducing poverty, and corporate governance. The fundamental idea behind the concept is to provide the same and better access to the social resources to the future generation and also to the generation of the today.

> It is the ability to function continually at a level of social wellbeing and harmony. Issues as war, poverty, injustice, and low level of education are signs of a socially unstable system. At an individual level, one should have the access to health care, nutrition, shelter, and education, in addition to cultural expression (Adams 2006; Basiago 1999; Mackenzie 2004).

(ii) Environmental

Environmental sustainability focuses towards a defined level of environmental quality and keeping the natural resource capital intact. The environmental sustainability supports matters such as renewable energy, cutting fossil fuel consumption and

---

[1]The 3 pillars of sustainable development. http://www.knowledge-pills.com/comkp1/kp/series/0 18_updated/002_company_sustainable_development/wha02/02wha02.htm. Accessed March 12, 2018.

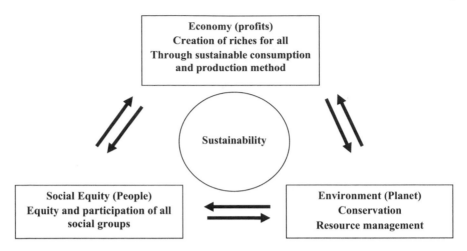

Fig. 2 Fundamental aspects of sustainability

emissions, sustainable agriculture, reducing deforestation, restoration of natural resources, recycling, and better waste control. The main objective of the environmental sustainability is the utilization of the renewable resources should not exceed the level at which it is rehabilitated, also the capacity of the environment absorbing waste should not be exceeded (Adams 2006; Basiago 1999; Mackenzie 2004).

(iii) Economical

Economic sustainability occur when development is financially feasible, whilst socially and environmentally sustainable (Gilbert et al. 1996). The part mainly focuses on internal and external parts of sustainability. This means a firm must also account the financial performance along with or by considering the social and environmental aspects (Adams 2006). Researchers suggest that the sustainable economy will give opportunity to measure the economic growth with the help of wide range of indicators such as investment, interest rates, productivity, and labour market and employment statistics over the only indicator gross domestic product value. The interactions between these factors should show whether the current levels of economic activity are sustainable (Doane and MacGillivray 2001). The triple bottom line approach suggests that companies should consider social and environmental performance, not only financial performance, in their business operations.

3 Fast Fashion and Unsustainability

The process of turning raw materials into finished garments has significant negative environmental and social implications, including air and water pollution and exploitation of human resources, especially where production is outsourced to lower labor cost countries (Shen et al. 2012). The textile and clothing industry is considered as one of the most polluting industries in the world (Austgulen 2015). The fast is defined as,

> Fast fashion—low-cost clothing collections based on current, high-cost luxury fashion trends—is, by its very nature, a fast-response system that encourages disposability (Fletcher 2008).

This system reduced the standard turnaround time of the apparel industry from six months to few weeks by adapting fast manufacturing cycles like, rapid prototyping, large variety, efficient transport and delivery. Leading mass retailers like H&M and Zara were reduced their cycle time and made new styles available in stores every month (Tokatli 2008). The increased attraction towards the fast fashion among the young customers is mainly associated with their higher credit availability and disposal income. The fast fashion system taps exactly this segment and encourages the customer to make frequent purchase and dispose immediately. This Fast fashion has been referred to as "McFashion," because of the speed with which gratification is provided. The framework is global, and the term "McFashion" is, to a degree, appropriate (Joy et al. 2012).

As the fast fashion industry grows, the utilization of the clothing material is increased considerably around the world. A research report by Ellen Mc Arthur foundation confirmed that the clothing utilization increased rapidly in the recent year up to 15% than the last years. The increased utilization of clothing observed in all the nations invariable to the economic nature of the country.[2] When compared to the previous year (2016), the total value of the apparel market increased to 842.7 billion USD with the hike rate of 5.5%. The Asia-Pacific alone contributes around 60.7% of the textile market value in the year 2016. The market report also forecasted that the market value will increase to 1004.6 billion USD in 2021. This is approximately 19.2% of the market growth when compared to the market value of 2016. A survey conducted by market line, a business information company depicted that, since 2011, the global apparel industry growing at a rate of 4.78% yearly and the value of sales is around 1.4 trillion USD for 2017. The research also projected a growth of 5.91% for the next three years for the apparel industry sector. They had also mentioned that at 2020, the market size will reach a value of 1.65 Trillion USD worldwide (Singh 2018). The fast fashion strategy mainly focuses towards fast consumption of the clothing by following the fast changing trends. The concept doesn't bother about the quality aspects of the apparels. Globally each year, millions of garments end up in landfill. Annually, the customer wastes around 460 billion worth of cloths approximately as waste around the globe.[3] The Ellen Macarthur foundation research mentioned that, some of the wasted garments discarded after just ten uses. The research quoted that, in the last fifteen years, the clothing sales increased double the time from 50 billion units in 2000 to more than 100 billion units in 2015. At the same time they have

---

[2]Circular fibres initiative analysis based on euro monitor international apparel & footwear, 2016 edition (volume sales trends 2005–2015).

[3]Circular fibres initiative materials flow analysis and euro monitor international apparel & footwear, 2016 edition (volume sales trends 2005–2015).

also mentioned that in world wide the usage percentage of clothing material reduced significantly than that of the previous year's (Ellen MacArthur Foundation 2017).

Environmental issues are the major problem, which can be expected out of the implementation of fast fashion system and the characteristic change of the society. To produce cheap quality product or less cost product, the manufacturer's uses different types of synthetic products in the manufacturing phase of the product. This potentially harms the environment. Also the cheat quality materials used in the manufacturing makes the apparels as a shot life product and which will ultimately ends in the landfill. The fast fashion system mostly uses synthetic fibers like polyester, polypropylene and etc., as a major source of fiber over cotton. Higher the utilization of synthetic fiber leads higher environmental impact. In case of polyester, the process sequence releases almost three times more carbon dioxide to the environment than the cotton. Polyester is one of the major synthetic fibers, which is present in almost all fast fashion items and represents 60% of the total clothing used. The deposits of polyester in the landfill, spoils the environment further (Cheeseman 2016). In 2017, around 21.3 million tons of polyesters used in the clothing production, which is approximately 157% higher than that of year 2000 (Cheeseman 2016).

From the year 1992 to 2002, the life of the consumer products are reduced by 50% as an impact of fast fashion system. The fast fashion industry consumes non-renewable resources like oil to produce synthetic fibres, fertilizers to grow cotton, and chemicals to produce, dye, and finish fibres and textiles (Watson 2016). Generally, all the fabric manufacturing processes are highly water intensive. The total greenhouse gas emission from the apparel and textile industry is around 1.2 billion tones, which is more than the total all international flight and maritime shipping emission (International Energy Agency 2016). Globally 20% of the water polluted by the textile industry like dyeing and finishing textile. Other than this, more than half million tones of plastic microfibers were let into ocean by the textile industry. (Kant 2012; O'Connor 2017). It is also estimated that the resource consumption of the textile industry will increase multiple times in the next few decades. The CO_2 emission will increase 26% in 2050 from 2% in 2015 (Leonard 2016).

4 Consumer Behaviour

Understanding the requirements of the customer is a very complex task for any manufacturer. Various researchers had employed different methods to understand the customers behavior by different means (Bagozzi et al. 2002; Simonson et al. 2001; Gabbott and Hogg 2016).

The common definition of consumer behaviour according to Perner is (Perner 2010)

> the study of individuals, groups, or organizations and the processes they use to select, secure, use, and dispose of products, services, experiences, or ideas to satisfy needs and the impacts of these activities on the consumer and society

Fig. 3 Purchase activity
flow as mentioned by
Hawkins and Mothersbaugh

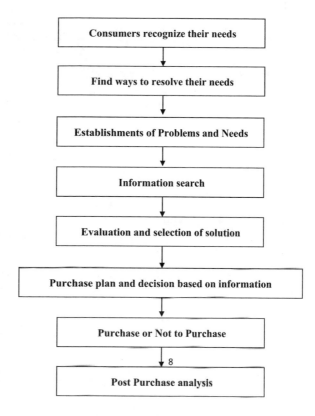

Every customer has three different role as user, payer and buyer (Furaiji et al. 2012). All this three activities are the part of the purchasing process of decision making process regarding a purchase. The activity flow was provided as follows by Hawkins and Mothersbaugh (Hawkins and Mothersbaugh 2010) (Fig. 3).

Other researcher classifies the customer's decision making process into different categories based on the product the customer purchases (Furaiji et al. 2012). They are,

(i) Normal Purchase behavior: Routine and normal low cost product which brought frequently. Eg: Groceries, Apparels.

(ii) Limited decision making: It is the purchase process when an extensive purchase decision is combined with a normal routine purchase. This is kind of situation where the customer go for a specific brand for a normal product.

(iii) Complex decision making: It is the most extended decision making. In this process the customer purchases very infrequent products and more expensive ones. Eg: Car, Big home appliances etc.

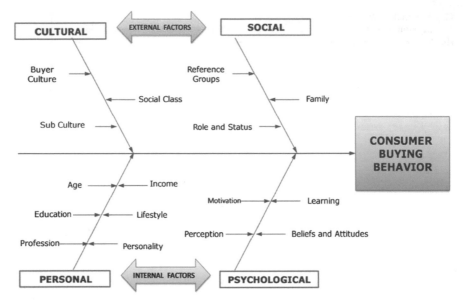

Fig. 4 Internal and external factors affecting consumers buying behaviour (Al-Salamin and Al-Hassan 2016)

4.1 Factors Influencing the Consumer Decision

Different researchers worked on various methods to understand the behaviors of the consumer and also the factors influencing the decision making process of the consumers either by individually or by collectively as a group based on some grouping factor (Hawkins and Mothersbaugh 2010). The consumer behavior can be influenced by two categories, namely Internal and external factors. Internal factor refers psychological factors and external one refers the sociological matters regarding the decision making skills of the customer. The Sociological factors can be like, culture, subcultures, demographics, social status, reference groups, family and marketing activities. Internal factors could be perception, learning, memory, motives, personality, emotions and attitudes (Hawkins and Mothersbaugh 2010) (Fig. 4).

4.1.1 Internal Factors

The main internal factor which influences the purchase is motivations. To properly understand the behaviors of the customer it is also important to understand the factor that motivates the customer. A motive is an inner urge (or need) that moves a person to take purchase action to satisfy two kinds of wants viz. core wants and secondary wants. So, motivation is the force that activates goal-oriented behaviour. Motivation acts as a driving force that impels an individual to take action to satisfy his needs.

So it becomes one of the important internal factors influencing consumer behaviour. The motivation happens while a need or want recognized by the customer. Once the need is identified, a state of tension appears that drives the consumer to attempt to reduce or eliminate the need. This need may be a functional/practical need, or it may be an experiential need, which involves emotional responses or fantasies (Solomon et al. 2006).

Perception is another important factor which influencing the internal behaviors of the customer. Human beings have considerably more than five senses. Apart from the basic five (touch, taste, smell, sight, hearing) there are senses of direction, the sense of balance, a clear knowledge of which way is down, and so forth. Each sense is feeding information to the brain constantly. Researchers mentioned that most of the time the customer also purchases the product for their values instead of the function of the product. Sometimes based on the perception, the customer feels that the products value is important and more than the primary function of the product. In general, customer wants more than the desired function of a product, they look at the image of the product, image of the company and after-sale service (Furaiji et al. 2012).

Researchers also mentioned that the generated customer need creates a tension or increases the level of tension in customer mind (Solomon et al. 2006). The goal of the customer is to reduce the tension to the normal level or to removes the tension. This is the main driving factor for the customer's internal choices and preferences. It is also important to understand that there will be always an exceptional behaviour among the customers. Researchers suggest that the same consumption behaviour of one customer is different at different times and occasions based on their need and so there is no standard rule, only guidelines (Hawkins and Mothersbaugh 2010).

4.1.2 External Factors

External influences are aptly named so because the source of influence comes from outside the person rather than from the inside. These factors include culture, social class, reference group and family influences. They are associated with the groups that the individual belongs to and interacts with.

(i) *Culture*

Culture is a major factor that influences consumers. Culture can be defined as (Hawkins and Mothersbaugh 2010):

> Culture is the complex whole that includes knowledge, belief, art, law, morals, customs, and any other capabilities and habits acquired by humans as members of society

Culture meaning that a complex sum total of knowledge, belief, traditions, customs, art, moral law or any other habit acquired by people as members of society. It is the values and a basic attitudes of the society with in which an individual flows that. Cultural norms are learnt by an individual from childhood and their influence is so ingrained that it is invisible in daily behaviour. Culture teaches an individual the acceptable norms of behaviour and tells him the rights and wrongs.[4] Within a culture, there are many groups or segments of people with distinct customs, traditions and behaviour. In the Indian culture itself, we have many subcultures, the culture of the South, the North, East and the West. Hindu culture, Muslim culture, Hindus of the South differs in culture from the Hindus of the North and so on. Products are designed to suit a target group of customers which have similar cultural background and are homogeneous in many respects.[5]

The culture is the factor which consolidates everything that influences the customer's behaviors from the external world by affecting an individual's thought process. Hawkins and Mothersbaugh mentioned that culture does not have a proper description or either a definition of rules instead it is the very nature how a majority of the people in a group are act and think. Often people not aware of the culture but by nature they act as such because it is natural (Hawkins and Mothersbaugh 2010).

(ii) *Social class*

Social class alludes to hierarchical arrangement of the general public into different divisions, every one of which implies signifies social status or standing. Social class is a vital determinant of consumer behaviour as it influences utilization pattern, way of life, media pattern, activities and interests of purchasers. The word social class represents the gathering of individuals who share level with positions in a general public. Social class is characterized by parameters like wage, training, occupation, and so on. For example, two purchasers gaining a similar pay may vary impressively in way of life when one has proficient capability at the post graduate level and is utilized at the senior administration framework of a multinational company, while the other is independently employed, with training kept to a couple of years of tutoring. Inclinations with respect to item and brand buys, media utilization designs, interests in quest for different relaxation time exercises change a ton among these two customers (see Footnote 5).

Researchers usually classify the consumer's social status into three categories namely upper class, middle class and lower class (Diamond and Diamond 2013). The consumption behaviors of these three class consumers are totally different because of their level of income. Higher the disposable income higher will be the spending. A person with more discretionary income usually referred as the higher level in the social class. This higher income motivates them to more purchases. However it does not mean that they buy more. If a consumer has had a lot of money all of his/her life, the purchase behavior tends to be more conservative and qualitative (Diamond and Diamond 2013). But the researcher also mentioned that the newly rich customers in the upper class have a completely different purchasing pattern, they buy more and less carefully. But, the middle class customer still have enough money to buy product. However they will be more careful on the purchase and not to spend money

---

[4]External factors that influences consumer behavior. http://www.yourarticlelibrary.com/consumers/3-external-factors-that-influences-consumer-behavior/12908. Accessed April 10, 2018.

[5]External environmental factors affecting consumer behaviour. https://www.wisdomjobs.com/e-university/consumer-behaviour-tutorial-94/external-environmental-factors-affecting-consumer-behaviour-10475.html. Accessed April 10, 2018.

on unnecessary things. In the lower class, consumers are very price conscious and have less to none discretionary income and also they are expected to spend only based on the essential needs with vital requirements.

(iii) *Reference group*

Reference groups are gatherings of individuals that impact a person's state of mind or conduct. A group is an accumulation of people who share some customer relationship, states of mind and have a similar intrigue. Such gatherings are pervasive in social orders. These groups could be essential where interaction happens every now and then and, comprises of family gatherings. Secondary groups are an accumulation of people where relationship is more formal and less individual in nature (see Footnote 4). People utilize these groups as reference points for learning attitudes, beliefs and behaviour, and adapt these in their life. Family and dear companions are considered to be primary reference groups in an individual's life due to their frequency of interaction with the individual and supremacy of these groups influence in a person's life. Classmates, neighborhood, associates, different colleagues are a piece of the secondary reference groups of a person. These could be political groups, work groups and study groups, benefit associations like the Lions, Rotary, and so forth. The conduct of a group is affected by other individual from the gathering. An individual can be an individual from different groups and can have changed impacts by various individuals from groups in his utilization conduct (see Footnote 5).

(iv) *Family Influence*

As mentioned in the previous section, the family is the most imperative of the essential groups and is the strongest source of influence on consumer behaviour. The family convention and traditions are learnt by youngsters, and they soak up numerous behavioral examples from their relatives, both intentionally and unintentionally. These behaviour patterns turn into a part of kids' lives. In a joint family, numerous choices are together established which additionally leave a connection on the individuals from the family. Nowadays the structure of the family is changing and individuals are going in additional for core families which comprises of parent, and ward kids. The other sort of family is the joint family where mother, father, grandparents and relatives likewise living respectively (see Footnote 4).

5 Consumer Behavior in the Fashion Industry

Buying a fashion product or an apparel product is mostly a gender based activity. Irrespective of the gender the choice of the cloths and preferences can be varied from an individual to another according to the color, brand, fashion, and material. For example, many people use color and contrast to express of feelings in their mind. In addition to that, factors such as product properties, designs, comfort and individuality are playing a decisive role on consumer's apparel buying decisions (Pereira et al. 2010). Consumers from different industries and section behaves different manners, especially in fashion industry, the behaviour of the consumer is a complex one and has wide influences from both internal and external factors as well. Throughout the history many companies and manufacturers tried to understand the costumer behaviour by different methods and processes. As mentioned in the previous section it is important to understand the fact that what motivates the customer clearly to understand the behaviour of the customer in better manner (Diamond and Diamond 2013). According to Strähle (Strähle 2017) fashion consumers often buy clothing due to the emotional need. According to history, many fashion consumers have selected clothing based on the name of the designer, which is emotionally motivated (Strähle 2017). For example, jeans used to be a practical garment used by workers and chosen rationally by consumers, but along came Calvin Klein and changed that fact, and the designer jeans market was born (Diamond and Diamond 2013). IN the contemporary society, clothing or fashion is identified as a way of expressing ones identify (McNeill and Moore 2015). To be in a group or society is one of the fundamental needs of a human. Hence, to be in a society, individuals are trying to adapt the values and social norms that society has implemented within that specific culture which also includes the fashion and apparels (Hofstede et al. 1991). Hogg and Abrams define identity as "people's concepts of who they are, of what sort of people they are, and how they relate to others" (Hogg and Abrams 1988). In fashion, most consumers want to express meanings about oneself and to create an identity. This is because identity is extremely important for fashion consumers and sometimes the factor can outweigh other important factors, such as being ethical, sustainable and functional (McNeill and Moore 2015).

But the other researchers (Birtwistle and Moore 2007) mentioned that the behavior of the customer depends on the lack of knowledge of the negative effect that the fashion industry has on the environment. Stahel also mentioned that the influence of socializing in the decision making process of the customer may often leads to over-consumption. As in the case of fashion industry, the fast fashion segment is the perfect example for this scenario (Strähle 2017). A study by Byun and Sternquist identified two strong factors which are related to the consumer behavioral responses. They are, short renewal cycle and a limited supply of goods. The researcher demonstrated that both the mentioned factors are strongly influencing the customer behavior. These factors encourage the customer's number of visits to a shop and also this affects the buying pattern of the customer (Byun and Sternquist 2008). In their research they have noted that if a product is unavailable or limited supply, they perceive that they are in need of it. The product becomes the consumers most desire requirement.

Another study reported that the male and female customers are having different behavior towards the apparel purchase (Workman and Studak 2006). The fundamental differences between the gender in clothing preferences like appearance and dressing resulted significant differences in the purchase behavior also. Their research concluded that men exhibited a need-based approach while women exhibited a want based approach in their decision making processes. As a result of this, the investigation differentiates between male and female consumers by seeking to investigate any differences that may exist between these two distinct segments. In general, the consumers perceive a positive relationship between the price and quality, in the case of apparel products (Boyle and Lathrop 2009). However, Miyazaki et al. (2005) high-lighted that the quality of the product analysed based on the intrinsic information (information about the physical product), extrinsic information (factors that do not make up the physical product). The research concludes that the relationship between price and quality holds in the presence of a secondary indicator. Other researchers analysed the consumer approach towards the fast fashion system and referred the system as "throwaway fashion" indicating the propensity for young consumers to purchase items of low cost and low quality with the intention of using them infre-quently and then disposing of them (Birtwistle and Moore 2007).

> The trend of throwaway fashion owes much to increases in fashion purchase frequency and a real reduction in price levels. Furthermore, fast fashion retailers, such as H&M, TopShop and Zara, sell garments that are expected to be used less than ten times at very comparative price points.

6 Methodology

6.1 Questionnaire Development

The survey was developed by adhering to the guidelines of the Global Reporting Initiative sustainability report. As per the report, the questioner was designed three broad categories like economic, environmental and social impact (GRI 2013). The survey was aimed to collect the combined attitude of the respondent toward the behavior of the clothing purchase, use and disposal attitude on the basis of above mentioned aspects (Sadiku 2014). The survey analyses the following categories,

1. Environmental Concern Analysis
2. Purchasing Behaviour Analysis
3. Sustainability knowledge Analysis
4. Social impact analysis
5. Carbon Footprint Awareness Analysis.

This analysis was performed using Google forms, an online free survey platform to attract more number of respondents. The target group of the respondent was mostly college students and also few recently passed out students who are employed in the professional organizations. The demographic data based on age group, sex, and the occupations of the respondent were obtained at the end of the survey.

6.2 Data Collection and Target Group

The target group of the study was consumers of varying age groups between 18 and 35 years of age without restriction to gender, ethnic background and profession. The

Table 1 Personal information of the respondents

Criteria	Frequency	Percent	Cumulative percent
Student	241	60.25	60.25
Professional	43	10.75	71
Employee	67	16.75	87.75
Self employed	29	7.25	95
Others	20	5	100
Total	400	100	

survey questionnaire was developed in Google forms and the link for the survey distributed among the college students of different discipline across the city. The questioners were also distributed among the working profession with educational background through emails and hardcopies in the city. The survey was conducted between, 1st November 2017 to 28th February 2018.

7 Results and Discussion

The survey conducted among a total number of 400 participants with a gender distribution of 72.1% female and 27.9% male. The major age group of the respondent was 15–35. Out of the selected group majority of the respondent are in the age group of 20–25. Personal information detail reveals that the majority of this survey respondent is undergraduate students. Out of 400 respondents, 241 respondents, 60.25% are students and the second major contributors are professional employees with the contribution of 27.25% (Table 1).

7.1 Environmental Concern Analysis

The main focus of this section is to study the consumer's attitude and concern towards the environmental sustainability and ethical consumption. To evaluate these characteristics the following questions were kept in the questionnaire. They are,

1. Do you think the apparel manufacturing process has considerable environmental impact?
2. Do you consider environmental impact of the clothing before purchase?

Question 1:

Out of the attended 400 respondent, 80.7% of the respondents were aware of the impact of the apparel manufacturing process or else they believe that the apparel products in the market are manufactured at the cost of environmental pollution.

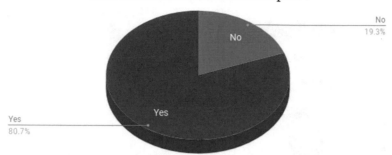

Do you think the apparel manufacturing process has considerable environmental impact ?

Fig. 5 Response for the Question number 1

Around 19.3% of the respondents only were not aware of the environmental impact of apparel manufacturing process. As the major respondents of the survey are students (60.25%), it can be understood that, most of the young generation people are aware of the environmental impact of textile and apparel manufacturing however, few of them are unaware of it (Fig. 5).

Question 2:

The result of second question details that, out of the total respondents, a sum of 149 respondents (37%) mentioned that they may consider the environmental impact of the clothing before purchase. About 46 (11.5%) of the respondent only said 'Yes' to the question and only 22 respondent, around 5.5% only said very often. Majority of the respondents (45.75%) said either No or rarely, around 183 members. From the results it can be considered that, around 45.75% of the respondent are not considering environmental factors before purchase and 37% of the people are in the middle state, where they will consider 'sometime' about all these factors. Only 17% of the people are really considering these factors during purchase of the apparels (Fig. 6).

From the result of these two questions it can be concluded that, even though the respondents know the impact of apparel manufacturing process on the environment, they were not considering the factor while purchasing new apparels.

7.2 Purchasing Behaviour Analysis

The focus of this section is to measure purchase behaviour of the respondent by observing how sustainability reflects in respondents purchase decisions and whether or not it is factored. The following questions were asked in the survey.

3. What kind of apparel you prefer to buy?
4. How often you will buy clothing and apparel?

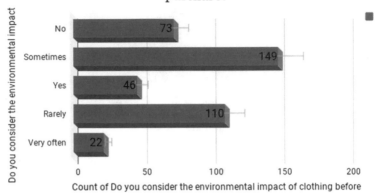

Fig. 6 Response for the Question number 2

5. What is the important factor you will consider while selecting apparel?
6. Where do you purchase your clothing usually?
7. Are you aware of environmental friendly clothing or clothing brands?

Question 3:

The results of question number three revealed that out of the 400 respondent, there is an equal preference for both durable and long life products and also for the style and trendy products. Among the respondent 158 members (39.5%) selected durable and long life products and 156 members (39%) selected the product that in trend. Very less percentage of the participants selected the fashionable products (13.25%), cheaper products (1.25%) and also the less fashionable products (6.75%) (Fig. 7).

Question 4:

This question is framed to evaluate the purchase frequency of the respondent. The results of the survey indicated that, most of the customers prefers purchase of new apparels once in a three month (167 out of 400; 41.7%). 31.5% of respondents (126) were also mentioned that they would like to purchase new apparels every month. This is the second major response next to the previous one. Very few percentage of the customers mentioned that they will buy only at the special occasions (17.75%) and once in a year (5.25%) (Fig. 8).

Question 5:

The majority of the respondent preferred 'Quality" as their choice for the question about the important factor which will be consider during the purchase. About 49.25% of the customers selected quality as their important factor in purchase. Next to quality 22.75% of the respondent mentioned that they prefer style over quality in the deciding factor during purchase. Few customers' decision also depends upon the

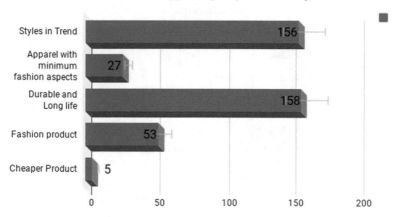

Fig. 7 Response for the Question number 3

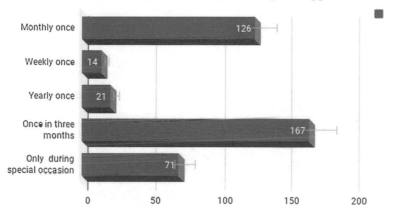

Fig. 8 Response for the Question number 4

price (15.75%). Surprisingly only 14 respondents out of 400 (3.5%) mentioned that they will consider environmental impact of the clothing material before purchase. This was in contrary to the first question, where the respondents were asked about the knowledge on the impact of apparel manufacturing process. Where 80.7% of the respondents mentioned they are aware of environmental impact caused by apparel production methods (Fig. 9).

Question 6:

In the case purchase location most of the customers responded as they will purchase the cloths anywhere. Out of 400 respondents, 160 respondents (40%) were ready to by anywhere if the cloths available as per requirements. Around 116 par-

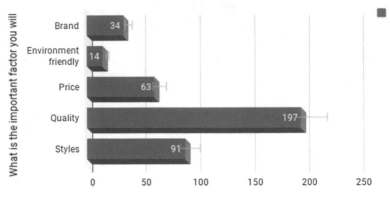

Fig. 9 Response for the Question number 5

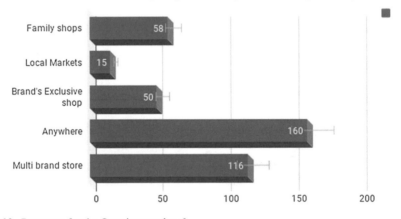

Fig. 10 Response for the Question number 6

ticipants (29%) mentioned that they used to purchase mostly from the multi-brand outlets. Only about 50 and 58 respondents preferred brand exclusive shops and family stores respectively. However, the percentage of the respondent preferred to go for brand exclusive shops and family shops are very less respectively 12.5 and 14.5% (Fig. 10).

Question 7:

This question is designed to understand the respondent's knowledge on the environmental friendly apparel brands. The response for this question makes clear that out of the total respondents 75.7% of customers are aware of the environmental friendly clothing brands. Only 24.3% of the respondents were not aware of the envi-

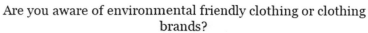

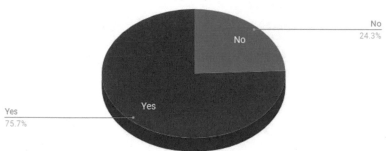

Fig. 11 Response for the Question number 7

ronmentally safe brands. Surprisingly, majority of the respondent said they are aware (Fig. 11).

7.3   Result Analysis

In this category, the result obtained from the respondents shows the very natural behavior of the normal customer. The respondents gave importance to the both quality and style during the purchase. Both the factors considered by equal number of respondent. Among the total respondent more than 75% agreed that they know about eco friendly clothing and brands. However, when the purchase comes into picture the behaviors of the customers noticed in contradiction.

The majority of the respondent mentioned they will buy either once in a month or once in a three month. Which indicates that the purchase behavior of the young customer not only by the needs and wants but by motivation by the other external factors. It is also to be noted that the customers mentioned that they prefers to monitor the product only in the aspect of the quality and not by the environmental impact. This point was further supported by the next questions, where the respondent mentioned that they will by the product anywhere if it is affordable, cheap and up to the expected quality level.

7.4   Sustainability Knowledge Analysis

8. Do you make any effort in buying environment friendly clothing?
9. Do you accept to pay more money for environmental friendly apparels?
10. How much are you ready to spend extra on your environment friendly apparel?

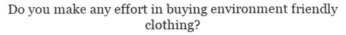

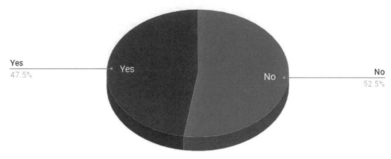

Fig. 12 Response for the Question number 8

11. How many clothing material/Apparel you have in your wardrobe?
12. Do you have unused apparels/clothing in your wardrobe?
13. For an apparel to be used for long time with you, what will be the most influencing factor?

Question 8:

This question is designed to understand the preferences of the respondent towards the environmental friendly product. Considering the environmental impact during the purchase should be the personal motivation during the purchase process. More than half of the respondents (52.5%) mentioned that they won't make any extra effort in selecting the environment friendly clothing. However remaining 47.5% said they will make extra effort to get eco friendly clothing (Fig. 12).

Question 9:

But it also noted that majority of the respondent, 65.3% are ready to pay more amount of for environmental friendly apparels. However, 34.8% of the customers mentioned that they don't want to pay extra money for the apparels even if it is environmental friendly (Fig. 13).

Question 10:

The question is focused to analyze the customer's intention towards the environmental friendly clothing items. From the answers it can be noted that out of the 400 respondents around 37.25% are ready to pay more than 25% of the actual price, 26% of the respondent mentioned they are ready to pay 50% more for environmental friendly products. In total, 81.5% of the customers are willing to pay more than 25% of the actual cost of the price. Around 18% of the customer are not interested towards the extra payment for the environmental friendly apparels (Fig. 14).

Fig. 13 Response for the Question number 9

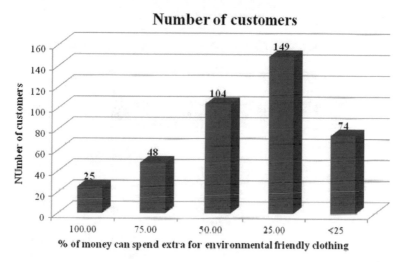

Fig. 14 Response for the Question number 10

Question 11:

For the question about the number apparels in the wardrobe 116 respondents mentioned that they have more number of cloths than they required (29%). 37 participants mentioned that they have lot of cloths then they actually need (9.27%). However, 206 respondents (51.5%) mentioned that they have only enough number of apparels an individual needs (Fig. 15).

Question 12:

Out of the participated respondent except 13.5%, all others having unused apparels in their wardrobe. In this section 99 respondents mentioned (24.75%) that they have one or two apparels as excess in their wardrobe, 152 respondents (38%) told they

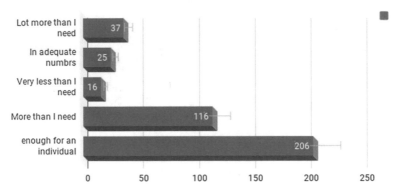

Fig. 15 Response for the Question number 11

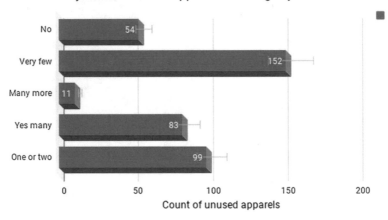

Fig. 16 Response for the Question number 12

have very few items as un used. However, 23.5% mentioned that they have many apparels in their cupboard as unused (Fig. 16).

Question 13:

The question designed towards the analysis of the most important intrinsic and extrinsic factor which influences the life time of the apparel product. For this question most of the respondent represented s quality is most influencing factor. Next to that the customers preferred design feature or style as their secondary factor (19.5%). However, the functionality, personal value and emotional value towards the apparel also considered important for some customers (Fig. 17).

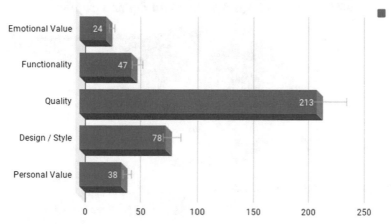

Fig. 17 Response for the Question number 13

7.5 Result Analysis

The abovementioned questions are designed to evaluate the customer's knowledge related to the sustainability aspects. From the result it can be understood that customers believed quality is the major factor for the long life of apparel product. The design and style is considered as a secondary factor. However, out of the respondent's majority of the consumers mentioned that they are conscious on the apparel product which are basically produced with lesser impact on environment. More than half of the respondent mentioned that they are ready to pay extra money for the eco friendly apparels. They even mentioned that they are ready to pay 25–100% extra price for the eco friendly apparels than that of their original price.

Even though the customers mentioned that they focus towards the environmental friendly apparels. However, individually the respondent mentioned that they have more apparel in their wardrobe than their requirements. From the analysis we can see that even though the customers are ready to spend for eco friendly apparels, they had more amounts of apparels in the wardrobe more than their requirements.

7.6 Social Impact Analysis

14. Do you believe that the usage of the clothing also has environmental impact during laundry and ironing?
15. How do you normally dispose your old cloths?
16. Are you aware of recycling process in used clothing?
17. Do you purchase and use second hand cloths or used cloths if offered?

Do you believe that the usage of the clothing also has
environmental impact during laundry and ironing ?

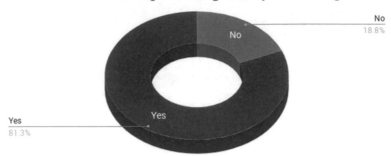

Fig. 18 Response for the Question number 14

18.   Are you interested in re-designing and extend the usage of your old cloths?

Question 14:

Out of the participants, 81.3% of the respondents positively mentioned that they
have knowledge on the laundering and ironing process and they believe these processes create considerable environmental impact as (Fig. 18).

Question 15:

Among the respondents, majority of the participants, around 58% of the respondent mentioned that they use to donate their old cloths to the charities. Hoe ever, while
asking about the charity process, they don have any clear idea about the activities of
the charity in terms of the collected used cloths. The respondents believe that, once it
is donated their responsibilities are getting over. Surprisingly around 26.25% of the
respondents mentioned that they will pass their old cloths to their family members.
However, 35 respondents (8.75%) discard their old apparels directly. Few respondents even mentioned that they will re design and use it for different other purpose
as (7%). It is also important to note that, none of the participants were aware of the
last option that, second hand shops for the used apparels. None of the respondent
selected that option (Fig. 19).

Question 16:

About 59.25% of the respondents were not aware of the recycling process. The
rest of 40.75% respondents either they have idea about the recycling process in details
or at least they are aware of the recycling activity in the industry (Fig. 20).

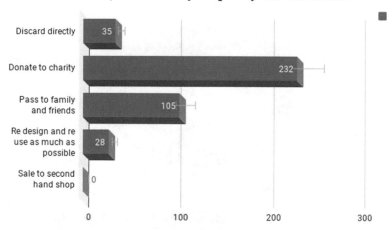

Fig. 19 Response for the Question number 15

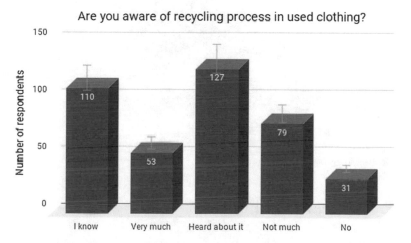

Fig. 20 Response for the Question number 16

Question 17:

Out of the 400 respondents, 87.5% of the respondents denied to use the secondhand cloths even if its offered in the market for the purchase. 12.5% of the total respondents are ready to purchase the second hand products if it is offered (Fig. 21).

Question 18:

Among the total respondents 77.4% of the customers mentioned that they are interested towards the re designing of their old cloths and extend the life of the product. However, a relatively few percentage (22.6%) are not interested towards the approach (Fig. 22).

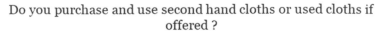

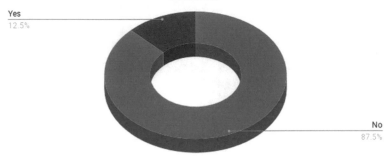

Fig. 21 Response for the Question number 17

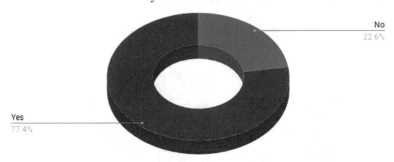

Fig. 22 Response for the Question number 18

7.7 *Result Analysis*

These questions are designed to analyze the knowledge of the respondent regarding the social impact of apparel utilization. As the questions are focused towards the increment of the apparel product life, questions about the apparel disposal and preferences were asked. Customers are responded that they follow the way to dispose their old apparels either by passing to their family members or by donating it to charity. When it is asked about the recycling, the respondents mentioned that most of them already know about it either or know about it as knowledge. The respondent also interested to use their old cloths if it is properly redesigned for same purpose or different purpose by up-cycling or down cycling process. However, when it is asked, whether you will by second hand clothing if it offered, most of the respondent said "No". This represents that the respondent's awareness towards the societal impact of the apparel product is limited.

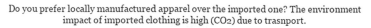

Do you prefer locally manufactured apparel over the imported one? The environment impact of imported clothing is high (CO2) due to trasnport.

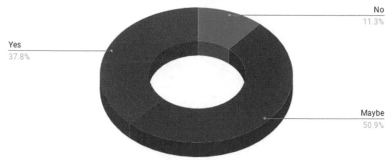

Fig. 23 Response for the Question number 19

7.8 Carbon Footprint Awareness Analysis

19. Do you prefer locally manufactured apparel over the imported one? The environment impact of imported clothing is high (CO_2) due to transport.
20. Will you prefer a brand that reduces the global climate change impact?
21. Are you aware of any fashion brand which is known as "sustainable brand" or "Environment friendly"?
22. Have you ever purchased "Environment friendly" clothing?

Question 19:

To understand the respondent's knowledge on the carbon foot print of the manufacturing process, the question has designed. From the result it is clear than more than 50% of the respondents were unaware of the process impact on the environment. Hence, 50.9% of the respondent mentioned that they may prefer if they had enough knowledge about the impact. About 37.8% percent of the customer directly accepted and they are ready to take the locally manufactured cloths to reduce the carbon foot print by this activity (Fig. 23).

Question 20:

Majority of the customers (60.9%) responded that they choose a brand with reduced environmental impact if they are educated about it. Very less, about 8% respondent said they will not change their purchase behavior at any cost. Altogether 92% of the customer preferred environmental friendly brand over their favorite brand, by considering the environmental impact of the ordinary branded product (Fig. 24).

Question 21:

This question is developed to understand the customer knowledge about the eco friendly brands in the market. A total of 54.3% customers are already aware of the ethical brand which had reduced 1or less environmental impact. However, nearly

Will you prefer a brand that reduces the global climate change
impact?

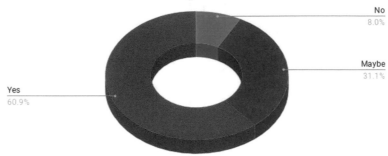

Fig. 24 Response for the Question number 20

Are you aware of any fashion brand which is known as "sustainable brand" or
"Environment friendly"

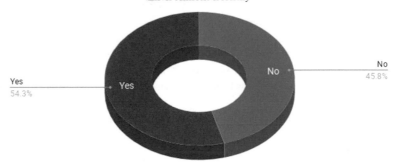

Fig. 25 Response for the Question number 21

half of the respondent, around 45.8% has no idea about the environmental friendly
brands in the market (Fig. 25).

Question 22:

It can be seen from the responses that majority of the respondents (61.3%) never
purchased the environmental friendly cloths. Only 38.8% of the respondent has purchased a product from the environmental friendly stores or brands (Fig. 26).

7.9   Result Analysis

The questions related to the carbon foot print analysis identified the customer's
knowledge about the environmental impact of manufacturing and transportation of
the apparel product. They customer mentioned that they may prefer locally manufactured product over the imported one if they have some idea about it. The respondent

Have you ever purchased "Environment friendly" clothing?

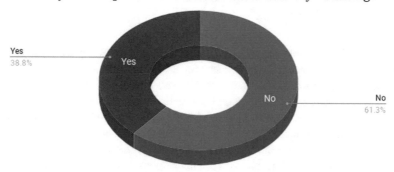

Fig. 26 Response for the Question number 22

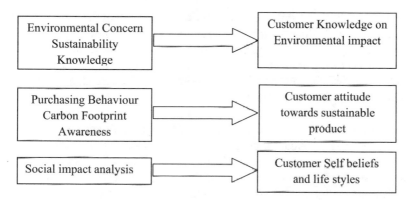

Fig. 27 Factors for the evaluation of the consumer survey

stated that they prefer eco friendly brands out of the normal brands. More than 54% of the customer agreed that they know eco friendly brands but in contrary to that 61.3% of the respondent mentioned that they never purchased any apparel from eco friendly brands before.

8 Discussions

The different sets of questions are framed under the different category to analyze the following behaviour of the consumers as mentioned in the Fig. 27.

8.1 Respondents Knowledge About Environmental Impact of Apparel Products

This section of questions was asked to assess the customer's key knowledge on the impact of apparel manufacturing process on the environment. Knowledge consists of the information stored within memory. The subset of total information relevant to consumers functioning in the marketplace is called consumer knowledge which will greatly affect individual's purchasing pattern. From the results, it can be understood that most of the respondents (80.2%) are aware of the impact of apparel manufacturing process on the environment. However when it asked "Will you consider the environmental impact before buying?", The responses were surprising that, more than half of the customers (45.75%) said they won't mind the factor during the purchase.

The result indicates that the respondents are very clear that the apparel manufacturing process has significant impact on the environment. But they are not considering this factor during the purchase process. This may be due to the non availability of the information on the product or due the lack of knowledge of the consumer on the product. These findings were in line with the findings of Kim and Damhorst (Kim and Damhorst 1998), who mentioned that although respondents had a moderately strong level of general environmental concern, this did not translate into environmentally responsible apparel consumption. Researchers are also mentioned that, it is not clear whether consumer's attitude or concern for the environment will always turn into positive environmental change (Troy 2007). The findings of a research conducted among 800 consumers stated that, the consumers believe that being green is an aspiration than the reality. So, the concern and knowledge of the environmental impact of the manufacturing process is not completely considered during the purchase of any product (Brosdahl and Carpenter 2010). A similar research finding were also reported by Borden and Schettino, who stated that a high level of concern about the environment does not consistently translate into a person actively seeking out information related to the product (Borden and Schettino 1979). Thogersen mentioned that (Thøgersen 2000)

> While consumers might be aware of the environmental impact associated with different behavior, they may be uncertain of the exact nature of the impact and thereby not understand the nature of the necessary behavior change. Further, while some consumers may be aware that a particular behavior is negative for the environment, they may not know how to change their behavior in order to be more environmentally sustainable.

Hence, from the results it can be concluded that knowledge is one important reason why consumers make unsustainable choices and that the more knowledge a consumer has about an environmental issue, the more likely the individual is to engage in environmentally preferable behavior (Thøgersen 2000).

8.2 Respondents Attitude Towards Environmental Friendly of Apparel Products

Attitude can be defined as simply and overall evaluation and it usually play a major role in shaping consumer behavior. In the normal situation the attitude is formed in two ways namely beliefs and feelings. These factors contribute in different manner at different situation or objects. This is one of the strong points which influence the behavior of an individual. Consequently, consumers' intentions to perform some behavior should increase as their attitudes become more favorable (Engel et al. 1995). In this analysis respondents mentioned that they prefer to purchase quality product and of course they bothered about the style and design aspect. The response to the question "Are you aware of Environmental friendly clothing and brand?" Most of the respondent mentioned that they have the knowledge about those factors. But when it comes to purchase, they are not sure about it. Even the customers mentioned that they are ready to spend extra money for the product. The respondents are also bothered about the impact of transportation and have a mindset to go for local brand. The results appeared to showcase the neutral behaviour of the customer even after the knowledge. This attitude of the customer is very common as referred by Butler and Francis (Butler and Francis 1997). The researchers mentioned that although consumers in the study held environmental attitudes which were more neutral in their attitudes about clothing purchase.

In general, the people are more concerned about the pollution in the earth and they believe that the pollution need to reduce and the environment must be protected. A survey conducted in the United Nations revealed that about 60% of the people in the survey have sound care about environmental issues and 49% believes the pollution level in the environment getting worse (Gallup 2013). The current results were very similar to the study conducted by Hustvedt and Dickson (Hustvedt and Bernard 2008). The current study respondent also disagree to buy eco friendly clothing, even they know the apparels are polluting the environment. This again suggests that the consumers are rather neutral in their attitudes towards sustainable clothing. The reason behind this neutral behaviour is very common among the general people due to their limited knowledge. One of the researcher mentioned that the general mind set of the respondent towards the environmental friendly organic cloths are; they are less stylish than the normal apparel in the trend, the customers believe that they are counter-culture in style, not well fitting, and generally uncomfortable to wear in the public (Connell 2011). Sometimes it also important for some customers to have a personal experience with the environmental issues like, pollution, environmental vulnerability and etc. This kind personal experience also will motivate the customers to change their attitude towards the utilization and consumption of the apparels (Dunlap and Jones 2002). Few other researchers statistically analyzed the customer's knowledge and their behaviors during purchase and they found either a very limited correlation (Hines et al. 1986) or no significant correlation between the knowledge and behaviour (Schahn and Holzer 1990).

8.3   Respondents Lifestyle and Beliefs Towards Environmental Friendly Apparel Products

Lifestyle is the patterns in which people live and spend time and money. It reflects a person's activities, interests, and opinions (Wang 2010). In this research most of the respondent had more number of clothing in their wardrobe more than their needs. This indicates their over purchased behaviour and less environmental concern. The respondents are also not aware of the possibilities of utilizing the unfit or old cloths. Most of the respondent mentioned that they donate unused cloths to the charity, however, when it is asked, the respondent have no idea about, what happen to the cloths after that?

This behaviour of the customers is purely not environmental friendly even though the responded that they have knowledge on apparels environmental impact and sustainable clothing, nothing reflects in their lifestyle. The respondent were mentioned that they will be glade if their old dresses are redesigned and offered again but at the same time they are refused to accept and use the second hand clothing. The major barrier for this kind of mid set their life style. When it is asked about the reason, the respondent mentioned that they are not sure about the cleanliness and they also worried about the acceptance among their most influencing groups like friends, family and colleagues. Mostly all the respondents mentioned that they are aware of the environmental brands and sustainable clothing names. But when it is asked more than 60% of the respondent mentioned that they never purchased their clothing from the eco friendly brands. This information clearly shows that the influence of life style nature among the customer. Consumers often choose products, services, and activities over others because they are associated with a certain lifestyle. So the lifestyle influence a consumer's purchasing behavior. During a study Tina Yinyin Wang noted that the profession, job profile and the nature of clothing that particular job profile required are all the factors which had great impact on the purchase and utilization behaviour of the respondent (Wang 2010).

9   Limitations of the Study

The study is focused towards the consumer behaviors towards the sustainable apparel purchase. However, the following points are considered as limitations of the study to generalize the results.

1. The survey conducted among 400 respondents in the Coimbatore city of Tamil Nadu, India. The results obtained may be different, if the sample size is larger and also the demographic influences were considered.
2. As the study mainly focused on college students and age group of 18–35, in this study the respondents income range was not considered. Grouping the respondents according to their earning capacity, gender, education status and the above mentioned age limit may also lead to different in the conclusion.

3. Some of the questions are designed with force choice question and neutrality, to achieve a reduced bias or social desirability among the respondents.
4. There are very limited number of questions were included in the aspect of responsible consumption of the apparel and about the circular economy concepts as this survey focused towards the knowledge, attitude and lifestyle of the consumer.
5. The survey didn't obtain any feedback from the respondent regarding the issues faced by the customers during the sustainable clothing purchase and utilization.
6. There are very limited numbers of questions asked to increase the interest of the respondent in filling up this survey.

10 Summary and Conclusions

The summary of this research work provides the noted behaviour of the consumer towards the sustainable apparels brands, purchase and utilization. The findings of the research are,

- Majority of the consumers told that they were aware of the environmental impact of the apparel manufacturing process but they are not interested towards the purchase of the environmental friendly product, they didn't even consider the environmental impact as a factor during the purchase. The customers do not have the moral attitude to engage in a sustainable and ethical purchase.
- The customer focuses any product in two aspects, one is quality, which they believe that will make the product long lasting and second one is style that's in trend. Out of these two factors, based on the motivating factors the importance of the customer varies time to time. This can be evident from their answer on purchase location, they said, they are ready to purchase the product anywhere if it is satisfies their need. They will not have any moral obligation towards the eco friendliness of the product at that stage.
- The attitude of the customer is not altered by their sustainable knowledge on the apparel and apparel manufacturing process they gained. The majority of the customers are willing to pay more than the required prices but due to their poor motivation factor the customers are not ready to search for the environmental friendly apparels. The intention and motivation by their internal and external factor are not influenced up to the expectation.
- Majority of the consumers said they are aware of sustainable clothing manufacturers, sustainable brands and they even ready to choose local brand if it offered to reduce the carbon foot print of the product. However, they were never purchased anything from the sustainable brands due to their basic attitude.
- The respondents are ready to accept the recycled and redesigned product from their own old cloths. But the respondents are not interested to use a second hand cloth if it is offered to them. This attitude of the customers clearly shows the influence of their life style and other groups in their purchase decision making process.

References

Adams, W. M. (2006). *The future of sustainability: re-thinking environment and development in the twenty-first century.* Report of the IUCN Renowned Thinkers Meeting, 29–31 January 2006. Available at: http://cmsdata.iucn.org/downloads/iucn_future_of_sustainability.pdf. Accessed March 15, 2018.

Al-Salamin, Hussain, & Al-Hassan, Eman. (2016). The impact of pricing on consumer buying behavior in SaudiArabia: Al-Hassa case study. *European Journal of Business and Management, 8*(12), 62–73.

Austgulen, M. H. (2015). Environmentally sustainable textile consumption—What characterizes the political textile consumers? *Journal of Consumer Policy.* https://doi.org/10.1007/s10603-015-9305-5.

Bagozzi, R., Gurhan-Canli, Z., & Priester, J. (2002). *The social psychology of consumer behaviour.* UK: McGraw-Hill Education.

Basiago, A. D. (1999). Economic, social, and environmental sustainability in development theory and urban planning practice. *The Environmentalist.* Available at: https://www.amherst.edu/system/files/media/0972/fulltext.pdf. Accessed March 12, 2018.

Birtwistle, G., & Moore, C. M. (2007). Fashion clothing—Where does it all end up? *International Journal of Retail & Distribution Management, 35,* 210–216.

Borden, R. J., & Schettino, A. P. (1979). Determinants of environmentally responsible behavior. *The Journal of Environmental Education, 10,* 35–39.

Boyle, P. J., & Lathrop, E. S. (2009). Are consumers' perceptions of price-quality relationships well calibrated? *International Journal of Consumer Studies, 33,* 58–63.

Brosdahl, D. J. C., & Carpenter, J. M. (2010). Consumer knowledge of the environmental impacts of textile and apparel production, concern for the environment, and environmentally friendly consumption behavior. *Journal of Textile and Apparel, Technology and Management, 6*(4).

Butler, S. M., & Francis, S. (1997). The effects of environmental attitudes on apparel purchasing behavior. *Clothing and Textiles Research Journal, 15,* 76–85.

Byun, S. E., & Sternquist, B. (2008). The antecedents of in-store hoarding: Measurement and application in the fast fashion retail environment. *The International Review of Retail, Distribution and Consumer Research, 18*(2), 133–147.

Cao, H., Chang, R., Kallal, J., Manalo, G., McCord, J., Shaw, J., et al. (2014). Adaptable apparel: A sustainable design solution for excess apparel consumption problem. *Journal of Fashion Marketing and Management, 18*(1), 52–69.

Cheeseman, G.-M. (2016). *The high environmental cost of fast fashion.* https://www.triplepundit.com/2016/12/high-environmental-cost-fast-fashion/. Accessed March 18, 2018.

Connell, K. Y. H. (2011). Exploring consumers' perceptions of eco-conscious apparel acquisition behaviors. *Social Responsibility Journal, 7,* 61–73.

Diamond, J., & Diamond, E. (2013). *The world of fashion* (5th ed.). New York: Fairchild/Bloomsbury.

Doane, D., & MacGillivray, A. (2001). *Economic sustainability the business of staying in business.* The SIGMA project. http://www.projectsigma.co.uk/RnDStreams/RD_economic_sustain.pdf. Accessed March 12, 2018.

Dunlap, R. E., & Jones, R. E. (2002). Environmental concern: Conceptual and measurement issues. In R. E. Dunlap & W. Michelson (Eds.), *Handbook of environmental sociology* (pp. 482–524). Westport: Greenwood Press.

Ellen MacArthur Foundation. (2017). *A new textiles economy: Redesigning fashion's future.* http://www.ellenmacarthurfoundation.org/publications. Accessed March 18, 2018.

Engel, J. F., Blackwell, R. D., & Miniard, P. W. (1995). *Consumer behavior,* 8th edn. The Dryden Press.

Fletcher, K. (2008). *Sustainable fashion and textiles: Design journeys.* London: Earthscan.

Furaiji, F., Latuszynska, M., & Wawrzyniak, A. (2012). An empirical study of the factors influencing consumer behavior in the electric appliances market. *Contemporary Economics, 6*(3), 76–86.

Gabbott, M., & Hogg, G. (2016). 7 consumer behaviour. *The Marketing Book*, 151.

Gallup. (2013). *Environment*. http://www.gallup.com/poll/1615/environment.aspx. Accessed April 10, 2018.

Gilbert, R., Stevenson, D., Girardet, H., & Stren, R. (1996). *Making cities work: The role of authorities in the urban environment*. London: Earthscan.

GRI. (2013). *Sustainability reporting guidelines* (pp. 3–94). Amsterdam, NY, USA: Global Reporting Initiative.

Hawkins, D. I., & Mothersbaugh, D. L. (2010). *Consumer behavior: Building marketing strategy*. New York: McGraw-Hill Irwin.

Hines, T., Bruce, M. (2007). *Fashion marketing: contemporary issues* (e-book). Amsterdam: Elsevier Ltd. Available at: http://www.academia.edu/6428586/Fashion_Marketing_Contempor ary_Issues_Second_edition. Accessed March 12, 2018.

Hines, J., Hungerford, H., & Tomera A. (1986/1987). Analysis and synthesis of research on environmental behavior: A meta-analysis. *Journal of Environmental Education, 18*(2), 1–8.

Hofstede, G., Hofstede, G. J., & Minkov, M. (1991). *Cultures and organizations: Software of the mind*. London: McGraw-Hil.

Hogg, M. A., & Abrams, D. (1988). *Social identifications: A Social Psychology of Intergroup Relations and Group Processes*. London: Routledge.

Hustvedt, G., & Bernard, J. C. (2008). Consumer willingness to pay for sustainable apparel: the influence of labelling for fibre origin and production methods. *International Journal of Consumer Studies, 32*, 491–498.

International Energy Agency. (2016). Energy, climate change & environment: 2016 insights (p. 113).

Joy, A., Sherry, J. F., Jr, Venkatesh, A., Wang, J., & Chan, R. (2012). Fast fashion, sustainability, and the ethical appeal of luxury brands. *Fashion Theory, 16*(3), 273–296.

Kant, R. (2012). Textile dyeing industry: An environmental hazard. *Natural Science, 4*(1), 23.

Kates, R. W., Parris, T. M., & Leiserowitz, A. (2018). What is sustainable development? Goals, indicators, values, and practice. *Environment: Science and Policy for Sustainable Development, 47*(3), 8–21. Retrieved from http://www.hks.harvard.edu/sustsci/ists/docs/whatisSD_env_kate s_0504.pdf. Accessed March 12, 2018.

Kim, H., & Damhorst, M. (1998). Environmental concern and apparel consumption. *Clothing and Textiles Research Journal, 16*(3), 126–133.

Leonard, G. H. (2016). *Oceans, microfibers and the outdoor industry: A leadership opportunity*. Presentation to Outdoor Industry Association (2016).

Mackenzie, S. (2004). *Social sustainability: Towards some definitions*. Hawke Research Institute. Available at: https://www.scribd.com/document/291037270/social-sustainability-towards-som e-definitions. Accessed March 12, 2018.

McNeill, L., & Moore, R. (2015). Sustainable fashion consumption and the fast fashion conundrum: fashionable consumers and attitudes to sustainability in clothing choice. *International Journal of Consumer Studies, 39*(3), 212–222.

Miyazaki, A. D., Grewal, D., & Goodstein, R. C. (2005). The effect of multiple extrinsic cues on quality perceptions: A matter of consistency. *Journal of Consumer Research, 32*, 146–153.

O'Connor, M. C. (2017). Inside the lonely fight against the biggest environmental problem you've never heard of. *The Guardian*, October 27, 2014. International Union for Conservation of Nature. *Primary microplastics in the oceans: A global evaluation of sources* (pp. 20–21).

Pal, R. (2016). Sustainable value generation through post-retail initiatives: An exploratory study of slow and fast fashion business. In S.S. Muthu & M.A. Gardetti (Eds.), *Green Fashion* (pp. 127–158). Singapore: Springer.

Pereira, M., Azeved, S. G., Ferreira, J., Migul, R. A., & Pedroso, V. (2010). The influence of personal factors on consumer buying behaviour in fashion. *International Journal of Management Cases, 12*(2), 509–518.

Perner, L. (2010). *Consumer behavior: The psychology of marketing*. Retrieved September 1, 2016, from http://www.consumerpsychologist.com/intro_Consumer_Behavior.html. Accessed March 12, 2018.

Rathinamoorthy, R. (2018). Sustainable apparel production from recycled fabric waste. In Muthu SS (Ed.), *Sustainable innovations in recycled textiles*. Singapore: Springer Nature.

Sadiku, O. O. (2014). *Sustainable and ethical fashion consumption: the role of consumer attitude and behaviour*. Master's Thesis, HSBA Hamburg School of Business Administration, Germany.

Schahn, J., & Holzer, E. (1990). Studies of individual environmental concern: The role of knowledge, gender and background variables. *Environment and Behavior, 22,* 767–786.

Shen, B., Wang, Y. L., Lo, K. Y., & Shum, M. (2012). The impact of ethical fashion on consumer purchase behavior. *Journal of Fashion Marketing and Management: An International Journal, 16,* 234–245.

Siegle, L. (2011). *To die for, Is fashion wearing out the world?*. London: Fourth Estate.

Simonson, I., Carmon, Z., Dhar, R., Drolet, A., & Nowlis, S. M. (2001). Consumer research: In search of identity. *Annual Review of Psychology, 52*(1), 249–275.

Singh, G. (2018) *Fast fashion has changed the industry and the economy*. https://fee.org/articles/fast-fashion-has-changed-the-industry-and-the-economy/. Accessed March 18, 2018.

Solomon, M., Bamossy, G., & Askegaard, S. (2006). *Consumer behavior a European perspective*. Harlow: Financial Times/Prentice Hall.

Strähle, J. (2017). Green fashion retail. In *Green Fashion Retail* (pp. 1–6). Singapore: Springer.

Thøgersen, J. (2000). Knowledge barriers to sustainable consumption. In P. F. Bone, K. R. France, & J. Wiener (Eds.), *Marketing and public policy conference proceedings* (pp. 29–39). Chicago: American Marketing Association.

Tokatli, N. (2008). Global sourcing insights from the clothing industry: the case of Zara, a fast fashion retailer. *Journal of Economic Geography, 8,* 21–38.

Troy, M. (2007). Panels: Consumer slow to go green. *Retailing Today, 46*(17), 24.

Wang, T. Y. (2010). Consumer *behavior characteristics in fast fashion*. Thesis, Boras, Sweden, August 2010. http://bada.hb.se/bitstream/2320/7723/2/2010.13.16.pdf. Accessed April 10, 2018.

Watson, D., et al. (2016). *Exports of Nordic used textiles: Fate, benefits and impacts* (p. 67).

Workman, J. E., & Studak, C. M. (2006). Fashion consumers and fashion problem recognition style. *International Journal of Consumer Studies, 30,* 75–84.

Motives of Sharing: Examining Participation in Fashion Reselling and Swapping Markets

Sarah Netter and Esben Rahbek Gjerdrum Pedersen

Abstract The aim of this chapter is to provide insight into the consumer motives to participate in new mobile-enabled reselling and swapping platforms. A netnographic analysis of app reviews from the U.S. iTunes app store is used to identify consumer motives for participating in mobile-enabled fashion sharing marketplaces. The findings from the study indicate that consumers are mainly driven by self-oriented motives (convenience, recreation, and product portfolio) whereas they are less inclined to occupy critical positions by considering reselling and swapping as a means to distance themselves from the conventional fashion industry. Within the last decade, the sharing economy phenomenon has gained momentum across the world. However, many initiatives are struggling to reach enough consumers and create a critical mass. Knowledge about consumer motivations may help sharing initiatives and policy-makers to strengthen the services and change consumer mindsets. Consumer behavior in the sharing economy has received limited scholarly attention. This exploratory study provides new insights into consumer motivations for adopting fashion reselling and swapping platforms.

Keywords Sharing · Swapping · Reselling · Sustainability · Consumer motivation · App reviews · Fashion

1 Introduction

Within the last decade, the Internet and mobile technology have given rise to a phenomenon that we have come to know as the sharing economy. At its core, the sharing economy encompasses a variety of different products and services, which

S. Netter (✉) · E. R. G. Pedersen
Department of Management, Society & Communication, Copenhagen Business School, 2000 Frederiksberg, Denmark
e-mail: sn.msc@cbs.dk

E. R. G. Pedersen
e-mail: ergp.msc@cbs.dk

© Springer Nature Singapore Pte Ltd. 2019
S. S. Muthu (ed.), *Sustainable Fashion: Consumer Awareness and Education* ,
Textile Science and Clothing Technology,
https://doi.org/10.1007/978-981-13-1262-5_2

are distributed or accessed by so called sharing practices. So far, there appears to be a lack of consensus of what actually constitutes '*sharing*' (e.g., Codagnone and Martens 2016), or rather what deserves the label of '*true sharing*' (i.e., Belk 2014b), considering the plethora of labels and concepts used interchangeably for bartering, gifting, lending, renting, reselling, sharing, and swapping activities, such as '*collaborative consumption*' (Botsman and Rogers 2010), '*(market mediated) access-based consumption*' (Bardhi and Eckhardt 2012; Belk 2014a), '*connected consumption*' (Dubois et al. 2014; Schor 2014; Schor and Fitzmaurice 2015), to name a few. Following the lead of the majority of policy-makers and regulatory bodies in referring to these emergent activities as '*sharing*', this chapter operationalizes the sharing economy as multi-sided platforms (B2C and C2C) that enable the compartmentalization of ownership and usership of goods, skills, and services by bringing together two or more distinct groups of users (Netter 2016). Popular examples range from short-term hospitality exchange, ridesharing, tool libraries, to fashion reselling and swapping platforms, to name a few.

Historically speaking, sharing practices have always been an integral part of everyday life, as in the case of circulating maternity and children's wear (e.g., Gregson and Beale 2004), furniture, or tools (e.g., Benson 2007). While so far, most of these practices have been performed on an informal level, i.e. within families, among neighbors or groups of friends, nowadays—fueled by the rise of the Internet and mobile technology—many of these former informal social practices are undergoing a transformation, facilitating sharing mobile, online and offline among former strangers. With the concept making its way out of the niche into the mainstream (Grimm and Kunze 2011; Seidl and Zahrnt 2012), the sharing economy is frequently championed as possessing disruptive potential to transform established industries and sectors (e.g. Walsh 2011), with sustainability being commonly considered a positive side-effect of sharing products and services (Botsman and Rogers 2010). Sustainability goals are claimed to be accomplished by means of reducing waste and use of resources, which would otherwise go into the making of new products (ibid.).

One product category that is assumed to hold enormous potential to contribute to a more sustainable development by being consumed in a more sharing manner is fashion items. With the rise of fast fashion retailers, offering trendy garments at shorter life cycles and low prices, a throwaway consumer attitude has emerged, with garments being discarded after few time wear (Birtwistle and Moore 2007). One way of breaking the cycle of the fast fashion business model is through extending the lifespan of unwanted clothes by means of redistribution and reuse instead of premature discarding (e.g., Lang et al. 2013). Reuse in this context is defined in line with the definition proposed by Morley et al. (2009), which conceptualizes reuse as the use of a product in its originally intended function.

With the growth of the sharing economy phenomenon, a number of business models have started to emerge, which cater to consumers who wish to redistribute their unwanted products and acquire pre-owned goods (Netter 2017). Mobile-enabled reselling and swapping platforms appear to be particularly promising, as they offer convenience and ease of access to a potentially large audience (Netter 2016). So far, however, there is a paucity of research on these redistribution markets. Considering the fact that most sharing initiatives are struggling to change consumer mindsets and behavior from throwaway consumption to sharing (Netter 2016), with fast fashion items often being used less than ten times before disposal (McAfee et al. 2004), gaining understanding of the factors that motivate consumers to participate in (presumably) more sustainable consumption ways is paramount to alter the sustainability profiles of the industry and consumers for the better. The aim of this chapter is therefor to provide initial insight into the consumer motives to participate in these new mobile-enabled reselling and swapping platforms.

So far, consumer behavior in the sharing economy has received limited attention. Exceptions include for instance Bardhi and Eckhardt (2012), who examine car sharing, and Zervas et al. (2015), investigating online reputation on Airbnb. There appears to be especially paucity of research when it comes to the sharing of fashion items. This study contributes to the existing literature by shedding light on the motives that drive consumers to participate in fashion sharing practices. By exploring this new avenue of research, insights will aid policy-makers and business leaders in promoting a more sustainable development in the fashion industry. The remainder of this chapter proceeds as follows: Sect. 2 provides an overview of the relevant literature, followed by Sect. 3, introducing the data and methodology as well as presenting the findings. Section 4 discusses the study results and Sect. 5 summarizes the conclusion and provides an account of implications and limitations.

2 Literature Review

While sustainable fashion consumption is a topic that only recently has gained more attention in the scientific discourse on sustainability (Hiller Connell and Sontag 2008; Markkula 2007), sustainable fashion consumption by means of redistribution has been studied to an even lesser extent, with the majority of studies focusing on recycling behavior and the decision when and how to part with unwanted clothing. Furthermore, most studies are of small-scale, exploratory nature. In light of this paucity of research, this section reviews literature on second-hand fashion shoppers' motives. More specifically, motives for discarding unwanted fashion items and acquiring preowned fashion items will be brought forward. As a first step, motives for parting with ones clothes will be presented. As a second step, factors that might drive consumers to acquire second-hand goods will be outlined.

2.1 Discarding Motives

Deciding which clothes to part with is not always a straight-forward task. Most often, it is a task that is not voluntarily initiated but brought on the consumer by external circumstances. Triggers for going through ones wardrobe are often seasonal, or, if not life-changing, than at least incisive events, like spring and autumn cleaning, moving in with one's partner, moving into a smaller apartment, or altogether emigrating with two suitcases in your hands (e.g., Laitala and Klepp 2011; Paden and Stell 2005). As Laitala and Klepp (2011) report, most consumers dispose of unwanted clothes twice a year in connection with the bi-annual rotation of clothes that is part of the spring and autumn cleaning, with women disposing of clothing more frequently than men.

Different reasons might drive or force consumers to part with some of their clothes. Based on the findings of her 2001 study, Klepp (2001) identifies five reasons for clothing redistribution: (1) technical or quality related reasons, (2) psychological reasons, (3) situational reasons, (4) "never worn" reasons, (5) functional reasons, and (6) sentimental reasons. Technical or quality related reasons for obsolescence pertain to the product not living up to the owners' expectations in terms of product performance (i.e., being ruined, worn out, or uncomfortable). Psychological reasons refer to the owner being tired of the product or style. Situational reasons can both relate to changed needs (i.e., body size, wardrobe composition, or living situation) or limited range of use. Products falling into the "never worn" category of obsolescence reasons are primarily impulse purchases or items received as a gift. In case of functional obsolescence, new and improved products have entered the market, which render the old item useless. Sentimental types of obsolescence pertain primarily to products being taken out of use and kept for other purposes or safekeeping. According to Klepp (2001), technical or quality reasons for obsolescence constitute the most common grounds for parting with unwanted pieces of clothing, closely followed by psychological and situational reasons.

According to Shim (1995), seven different patterns can be distinguished, based on the discarding method and the underlying motivation, i.e., (1) donation, (2) reuse, (3) disposal, (4) charity-, (5) environmentally-, (6) convenience-, or (7) economically-motivated act. Although clothing donations are frequently associated with altruism, Ha-Brookshire and Hodges (2009) found consumers to be rather driven by self-oriented motives, namely the wish to create more space in their wardrobes for new things, thereby closely linking clothing disposal to clothing acquisition. Although closet clearing frequently evoke feelings of guilt—i.e., over (repeated) purchase mistakes, overconsumption, or simply for not wearing an item enough—these feelings pass and make way for new shopping acquisitions. Besides feelings of guilt, the sorting through items that could potentially be discarded is also connected to feelings of anxiety, whether they are making the right choice or not (i.e., keep or give away). Ha-Brookshire and Hodges (ibid.) suggest that these feelings disappear once the unwanted items have left the household. Cleaning up ones closet thus results in the satisfaction of both utilitarian (achieving goal of cleaning up closet) and hedonic values (feeling better by reducing guilt and anxiety). Similar observations were made by Morgan and Birtwistle (2009), who identified that young consumers were especially plagued by feelings of guilt in case of expensive, higher quality clothing items that they had seldom or never worn. Donating these items to charity organizations made them feel better, thus provided a source of relief. Domina and Koch (1997a, b) also identified intrinsic rewards from helping people in need as a major motive for those who prefer to donate to charity organizations. Reuse or reselling via consignment shops or garage sales on the other hand were the most frequently used options for respondents driven by economic or environmental reasons for garments that were still perceived as valuable.

Hence, the economical or sentimental values of clothing do not only determine whether items should be redistributed but also through which channels. As Morgan and Birtwistle (2009) suggest, cheap fashion items were mainly disposed when the quality had declined, new fashion trends had become available or if items were bought for a single event. While cheap, low quality garments were simply discarded once they were considered unwearable, more expensive, higher quality items were primarily donated to charity organizations. Charity shops and hand-me-down to family and friends constituted the primary redistribution options chosen by respondents, with only few choosing reselling outlets (e.g. eBay, second-hand shops) or swapping events. Engaging in swapping events appears to be especially appealing for those, who are only interested in wearing fashionable items a few times. Similarly, Ha-Brookshire and Hodges (2009) report that most respondents preferred to give their unwanted items to family members and friends.

2.2 Acquisition Motives

Guiot and Roux (2010) propose three second-hand shopping motives, based on a study with 708 respondents in France, namely (1) critical motivations, (2) economic motivations, and (3) hedonic/recreational motivations. Although this study does not exclusively focus on the fashion context, it appears suitable in light of a lack of more tailored alternatives.

Critical motivations pertain to distancing oneself from the conventional market system, withdrawal from and grievance toward consumerist society, as well as embracing sustainability dimensions by focusing on ethics and ecology (i.e., through recycling and waste reduction). Similar findings were reported by Hiller Connell (2011), who suggests that purchasing clothes second-hand constitutes one form of patronage of environmentally friendly acquisition sources. Environmental considerations, such as reducing one's own footprint and waste going to landfills, are thus closely linked with this acquisition form.

Economic motivations encompass gratification from lower prices, the ability to afford more, as well as finding a fair price. Similar observations were made by O'Reilly et al. (1984), which is in line with the origin of second-hand consumption as a source of poverty alleviation (e.g., Williams 2002). According to Williams (ibid.), this has historically led to some sort of stigmatization, with second-hand clothes having been perceived as a sign of poverty.

Hedonic or recreational motives pertain to treasure hunting, originality of finds, social contact and interactions in the second-hand outlets, as well as nostalgic pleasure, deriving from the past, history and authenticity of a piece of garment. The thrill of the hunt, i.e. the hedonic elements of the unforeseeable shopping experience, and the possibilities of individualizing ones style also play a decisive role for some consumers according to Bardhi (2003) and Bardhi and Arnould (2005). Furthermore, finding something unexpected and unique is closely linked to matters of identity construction and demarcation. This is in particular true for fashion taking a special role providing a "second skin", an extension of the self that communicates to others (Belk 1988). Retro and vintage clothing offer the opportunity to more easily distinguish oneself from the mainstream and to develop a more personal style (e.g., Gladigau 2008). Second-hand channels as acquisition source furthermore provide a source of status and individuality, having the skill and expertise to make great finds (e.g., Roux and Guiot 2008). The importance of the social elements of the shopping encounter, whether at car boot sales, flea markets or other second-hand environments have further been highlighted by a number of studies, for instance Gregson and Crewe (1994), Crewe and Gregson (1998), and Belk et al. (1988).

3   Methodology

The study of consumers' motives for redistributing unwanted clothes is not without methodological challenges, as clothing disposal—especially by means of garbage disposal—can be considered a sensitive topic, with respondents being unwilling to indicate actual reasons for disposal, if these are deemed unseemly by social norms and standards. While qualitative approaches usually provide valuable insights into unchartered territory, the above mentioned methodological challenge becomes particularly apparent when personal focus group or face-to-face interviews are applied. As Laitala and Klepp (2011) report, only few respondents admit to throwing their old clothes into the garbage. In the same vein, only few female respondents admit that fashion aspects (e.g., out of style, bored with garment) are reasons for disposal.

In order to circumvent this problem, the study at adopts a netnographic approach for identifying the motives that drive consumer participation in mobile-enabled fashion sharing marketplaces. The netnographic method is a qualitative approach, made popular by Kozinets (2002), which applies ethnographic techniques to studying online consumer behavior. This approach is very beneficial, compared with traditional approaches, due to its use of natural occurring data. This is not only a rather time and cost efficient data collection approach, but also less obtrusive than face-to-face interviews, which take place in an artificial interview situation.

3.1   Data Collection

Data collection took place in September 2014 in the U.S. iTunes app store. Upon reviewing the variety of English speaking mobile-enabled fashion redistribution platforms, the decision was made to focus exclusively on the most mature app, i.e. the platform with the most user reviews. This anonym redistribution platform, hereafter referred to as iSHARE, allows both reselling and swapping of fashion items. All app reviews posted until September 5. 2014 were collected by making direct copies. The initial sample size of iSHARE app reviews was 631. The average length of the textual reviews was 196 characters, with the shortest review being comprised of 2 characters, the longest of 2137. The majority of reviews were shorter than 203 characters (67%). The majority of reviews provided information on motivation (57.53%), yielding an actual sample size of 363 usable responses.

3.2 Content Analysis

In order to identify the motives that drive consumers to use iSHARE, a systematic content analysis of the app review sample was carried out by means of using Microsoft Excel, version 2010, for frequencies and percentages. According to Braun and Clarke (2006), thematic analysis can be defined as "a method for identifying, analyzing and reporting patterns (themes) within data" (p. 79). The research team adopted an iterative coding process (Flick 2009). Upon repeated close reading of the raw data, initial key ideas and concepts which appeared critical for the study were identified and noted in a codebook. This process was repeated until no new themes emerged. Subsequently, the complete list of themes and ideas was explored, in order to collapse fitting themes together, and identify emerging patterns. The interpretation of the themes was accomplished through a repeated process of familiarizing and reading the material, as well as reviewing of relevant literature. The resulting themes were checked again against the data from the app reviews and were combined to nine codes and three overall categories. Subsequently, the research team searched for categories among these codes. These categories were checked to be mutually exclusive and clearly defined. The app review data was consequently analyzed, with any piece of text which related to one of the identified collapsed themes was marked and categorized under the corresponding number of codes. Room was given for the emergence of new codes during this process. As previous research on app reviews has shown (e.g. McIlroy et al. 2015), it is not possible to code for single themes, as most reviews are written in an unstructured and informal manner. Each theme was coded using a dichotomous scale ("1" mentioned, "0" not mentioned).

4 Findings

Table 1 shows the coding scheme emerging from the analysis. Three overall groups of motivations were identified, which each comprised of three themes. *Economic motivation*, *hedonic motivation*, and *functional motivation* constitute the overall groups. *Affordability*, *auxiliary income*, and *value for money* belong to *economic motivations*, *recreation*, *wardrobe update*, and *social component* constitute *hedonic motivations*, whereas *convenience*, *product portfolio*, and *safety* belong to the group of *functional motivations*.

Table 1 Motivations for sharing

Economic motivation (*driven by the idea to save or make money by using the platform*)	Hedonic motivation (*driven by the anticipated individual pleasure generated from using the platform*)	Functional motivation (*driven by the technical and functional aspects of the service offered by the platform*)
Affordability (*fashion sharing/reselling in order to save money*)	**Recreation** (*fashion sharing/reselling as a leisure activity*)	**Convenience** (*fashion sharing/reselling as a convenient means of acquiring and disposing pre-owned items*)
Keywords • *Cheap* • *For free* • *Sign-up credit* • *Save money* • *Great pricing*	*Keywords* • *Addicting* • *Bargain hunting* • *Thrill of the hunt* • *Buy a lot* • *Fun* • *Window shopping*	*Keywords* • *Easy* • *Simple* • *Efficient system to sell clothes/fast sale* • *Well organized platform*
Auxiliary income (*fashion sharing/reselling in order to make money*)	**Wardrobe update** (*fashion sharing/reselling as a means to remain fashionable*)	**Product portfolio** (*fashion sharing/reselling as attractive means of accessing wide variety of items*)
Keywords • *Make money* • *Extra cash*	*Keywords* • *Clean out closet (out of style, do not need/use anymore)* • *Expand wardrobe* • *Variety of new clothes and accessories* • *Individual style, inspiration*	*Keywords* • *Great range* • *Brand "XY"* • *Great quality*
Value for money (*fashion sharing/reselling in order to find great brands at a great price*)	**Social component** (*fashion sharing/reselling as a social experience*)	**Safety** (*fashion sharing/reselling as a safe means of acquiring and disposing pre-owned items*)
Keywords • *Great deals* • *"XY" brand for great price*	*Keywords* • *Community spirit* • *"Great girls"*	*Keywords* • *Feel safe* • *Protection of personal data* • *Protection of buyers (withhold money until transaction completed)*

To give some examples, *affordability*, an *economic motivation*, was coded for keywords indicating the ability to afford more items on these platforms, such as *great pricing, cheap, for free*, the platform *credit* reviewers received upon sign-up, which they could use to shop on the platform. *Recreation*, a *hedonic motivation*, covered such issues as *fun*, *window shopping*, the *addictive* element of reselling and swapping items via the studied platform, the "*shopaholic*" element, i.e. the ability to *buy a lot*, and the *thrill of bargain hunting*. *Convenience*, a *functional motivation*, pertains primarily to the *ease of use* of the platform. This is both in terms of *eased*

Table 2 Descriptive statistics

Type of motive	N	Percentage (%)
Functional motivation	**244**	**67.22**
Convenience	169	46.56
Product portfolio	92	25.34
Safety	19	5.23
Hedonic motivation	**214**	**58.95**
Recreation	107	29.48
Wardrobe update	78	21.49
Social element	73	20.11
Economic motivation	**143**	**39.39**
Affordability	76	20.94
Auxiliary income	53	14.60
Value for money	32	8.82

access to redistribution channels, the *organization* and *structure* of the platform, as well as the *efficiency* to *quickly sell* unwanted items.

Looking at Table 2, which reports the mentioning frequencies and percentages by type of motivation, it becomes apparent that *functional motivation* constitutes the most dominant driver, with 67.22% of the reviews reporting functional elements, such as the *convenience* of using reselling and swapping platforms, range and quality of the *product portfolio*, and *safety* issues. *Hedonic motives* were found to be relevant for more than half of the reviewers, with 58.95% of the reviews referring to *recreational* elements, the possibility to *update ones wardrobe*, and a *social* component of trading clothing and accessories online. The third type of motives, i.e. *economic motives*, were indicated to a lesser extent, namely by 39.39% of all reviewers. Economic motives pertain to the *affordability* of products offered on the platforms, the opportunity to generate *auxiliary income* by selling ones unused clothes and accessories, and a feeling of finding great deals, i.e. *value for money*.

Convenience was addressed by almost half of the sample (46.56%). *Convenience* in the context of this study pertains to the ease of use of the platform, how organized and structured it is, and how efficient the provider is at facilitating trades. It was mentioned by almost half of the sample (46.56%). One example is from the user *SalTomato*:

> Hooked! What a great forum for getting my shopping fix. I don't always have time to go to goodwill, no do they have everything organized by my size. I do have time to look through this app when I'm winding down my day and I can see clothes specific to my size and tastes. I recommend asking measurement questions or your stuck with reselling an unfit product. It's not hard. It's incredibly handy.

Recreation, mentioned by 29.48% of the reviews, pertains in the context of this study to the fun elements of using the platform and how addicting it is. As noted by one of the platform users *Eihort*:

> Addicted! I love this app. I can window shop any time I want, and you can't beat the
> convenience. Just in the first two days of using the app I bought 3 shirts, 2 dresses, and a
> killer pair of heels and still spent less than 80 bucks, shipping included! It's secure and easy,
> and I even got thank you cards or notes in the packages from the sellers. I hate having "in
> season" clothes and running into people with the same outfit, but with this app those days
> will be over! Love, love, love.

Furthermore, this dimension pertains to the ability of the provided service to satisfy
both ends of the shopping spectrum. While some use it primarily for window shopping, i.e. the satisfaction of the hedonic shopping need without having to purchase
anything, others value the platform for providing the opportunity to buy a lot and
satisfy ones shopaholic gene. Other reviewers, such as *Dance.wynner*, again express
the joy they experience from the thrill of the hunt, when hunting for bargains.

> In love. I'm so in love with this app, it took my shopping addiction to a whole new level.
> You can find amazing deals and really nice clothes. The only downside is the shipping is a
> tad expensive but other than that this app is amazing!

Product Portfolio in the context of the study at hand pertains to the range of product
items, the brands on offer, and the quality of the items offered for resale and swapping.
It was mentioned by 25.34% of the reviews. The following quotes by *RachelBeau*
and *LaLa123!* exemplify how appealing the wide variety of product can be:

> Love this app! Found so many edgy girly clothes! Been up all nite buying n selling, woo-hoo!
> [Name of the app] is an awesome place to shop! It's so convenient & an easy way to shop! It's
> so great to search around for your exact size & put awesome outfits together under different
> sellers! Great display of the things you want to buy! Prices are reasonably priced & shipping
> is great! I Like it so much; it's inspired me to be a seller too!

Wardrobe update was indicated by 21.49% of the reviews. In the context of this study,
wardrobe update pertains to cleaning out ones closet, as items are considered to be
out of style, do not fit anymore, or are simply no longer needed. It also pertains to
expanding and changing ones wardrobe, having access to a variety of new clothes
and accessories which suit ones individual style and preferences and can be used
as inspirational sources. The wardrobe update aspect is exemplified below with the
following quotes by the users *Succubusdesi* and *Sm_fry*:

> Love [this] app. Love [the name of the app] this app. Just had a baby so I have a lot of clothes
> that don't fit. Swapping clothes has been great. Now I have clothes that fit without having a
> pile that doesn't. […]
> Awesome – LOVING this! Really awesome app! I love the community feel, its totally
> something different. I have found some awesome clothes on there too and found an easy
> way to get rid of clothes I never wear! Really loving it!!!

Affordability is also a parameter when it comes to the offered products. Indicated by
20.94% of the reviews, this motive is concerned with the price-point of the platform
in relation to the available financial resources of the platform users. To give a few
examples by the users *Heather Grove* and *Hipgirl22*:

> Amazed. I love this app so much! Everything is super affordable and really cute. I'm very impressed.

> Love it. Love this app. Lots of great items and you can't beat the prices. Love shopping while making new friends.

Sharing also has a social aspect. More specifically, the social element of using the sharing platform, the community spirit, was indicated by 20.11% of the reviews. Two reviews by *Msnotforfoolishness* and *GGACA* may be used to illustrate this:

> This website is full of nice clothing at reasonable prices. Some of the things are so adorable and I have actually found a new friend on the website that is just wonderful!!!

> Amazing app. What's more fun than borrowing clothes and shoes or accessories from your best friend/ sister/ roommate? Not having to give them back and you getting paid for letting others play with your stuff! Super easy to use, great offerings…it will become your endless closet:)

The importance of social interaction in the shopping or swapping encounter has been highlighted by previous research (e.g. Belk et al. 1988; Gregson and Crewe 1994; Crewe and Gregson 1998). Similar findings have been reported by Sherry (1990), on the social embeddedness and experiential aspects of consumer behavior at a Midwestern American Flea Market. Likewise, Pedersen and Netter (2015) identified the social interaction in fashion libraries as a primary driver for consumers to sign up to and make use of these services. Interesting in this context is the work by Williams and Paddock (2003) and Williams (2003), who suggest that the search for sociality, distinction, discernment, and fun exist in parallel with economic necessity in people's reasons to engage in alternative consumption spaces.

5 Discussion: Functionality Over Criticality

Increasing reuse by means of sharing has tremendous potential of not only decreasing amount of clothing finding way on landfills and incineration without ever being worn but potentially also lowering the need for new clothing, therewith reducing the production of new clothing and resource extraction. In order to promote this development, it is crucial to understand the factors that drive consumers to change their behavior. While some might be motivated by altruistic considerations, such as saving or protecting the environment, others might be driven by rather hedonic motives.

Overall, this study found that consumers are motivated by *functional*, *hedonic*, and *economic* factors to use fashion reselling and swapping platforms. The findings are not without precedents in the academic literature. When it comes to functionality, previous studies have also highlighted the role of reliability, convenience, and accessibility for fashion consumption practices. For instance, Shaw et al. (2006), Niinimäki (2010), and Beard (2008) all suggest that inconvenience and discomfort constitute a major motivational barrier for consumers to pursuing more sustainable fashion consumption practices. These findings are in line with the ones reported by

Hiller Connell (2011), who suggests that the lack of organization of merchandise in second-hand shops and consumers' available time resources might prevent consumers from frequenting these sustainable outlets. Last, Domina and Koch (2002) suggest that convenience and access do not only determine the choice of a textile recycling channel but also have an impact on how frequently consumers make use of it, how much they dispose of and the variety of items.

Hedonic values have also been highlighting in previous research on shopping decisions. As Goldsmith et al. (1991) suggest young female fashion consumers appear to be strongly driven by fun and excitement in their fashion shopping behavior. Similarly, Bardhi (2003) found that hedonic elements, the thrill of the hunt, the unforeseeable nature of the shopping experience motivate consumers to choose second-hand shopping outlets. In a study on four Scandinavian fashion libraries, Pedersen and Netter (2015) found that library members highly valued the ability to play and experiment with outfits and items, without having to commit them to a (permanent) purchase. Last, Iwanow et al. (2005) suggest product quality and style constitute two of the main parameters influencing the general fashion purchasing behavior.

Economic factors have also been emphasized in previous research. As Hustvedt and Dickson (2009) report, financial resources and pricing constitute one of the main barriers, preventing consumers from adopting more sustainable fashion consumption practices. Similar findings were reported by Hiller Connell (2010) and Iwanow et al. (2005). However, Joergens (2006) suggests that consumers might not automatically act more sustainable if their financial resources allow it. Personal benefits, such as price, frequently outweigh less self-centered motives. If consumers have the financial resources to act more sustainable, they might not be inclined to do so, if this would entail being limited in the quantity of clothing they could otherwise acquire (ibid.).

A striking finding of this study is that none of the platform users indicated critical motives, such as distancing oneself from the conventional market system, consumerist society, and embracing sustainability. Looking at the second-hand literature, it becomes apparent that both disposal of unwanted clothes and acquisition of second-hand clothes are usually associated with critical motives, such as a wish to protect the environment (e.g. Shim 1995; Domina and Koch 1997a, b; Guiot and Roux 2010; Hiller Connell 2011). However, this does not seem to be the case in this study, where there is little indication of consumer activism. Previous research on sustainable consumption has shown that consumers' decision-making is largely constrained by external factors, frequently referred to as the "triple A", i.e. the availability, affordability, and accessibility of sustainable alternatives (Thøgersen 2010). As Hiller Connell (2010) suggest, even the most motivated consumers might be impeded to change their fashion consumption behavior to more sustainable practices, due to restraining external contextual factors.

However, despite the lack of indication of clear critical motives, the study at hand has shown that mobile-enabled reselling and swapping platforms might be a channel which will enable consumers to consume fashion in a more sustainable manner. The factors that usually hinder consumers from doing so, i.e. accessibility (i.e., convenience), availability (of an appealing product portfolio), and affordability, actually constitute the main factors explaining the use of these services. Therefore, these platforms may accidentally help to transform the consumption practices of those less concerned with the environment towards more sustainable practices. In marketing these mobile-enabled reselling and swapping platforms, providers are therefore well advised to focus less on the sustainable upsides of using these services, and rather highlight the self-oriented "triple A", i.e. the functional and hedonic benefits of using the service provides.

6 Conclusion

This study contributes to the emergent body of literature on the sharing economy by providing initial insight into the consumer motives for participating in new mobile-enabled fashion reselling and swapping platforms. So far, consumer behavior in the sharing economy has received limited attention. This is even more so when it comes to the fashion context. With the special role of fashion, providing a "second skin", an extension of the self that communicates to others (Belk 1988), it remains to be tested whether the identified factors motivating consumers to use mobile-enabled fashion reselling and swapping platforms are transferrable to other peer-to-peer sharing marketplaces.

With *convenience, recreation*, and *product portfolio* constituting the main motivating factors indicated by the platform users in their reviews, it can be deduced that these platform users are rather driven by self-oriented motives. More specifically, they appear to be especially driven by *functional motives*, i.e. the convenience associated with the service and the products on offer, followed by *hedonic motives*. *Economic motives*, such as affordability, the ability to make some extra money, and value for money, appear to have less importance. Contrary to previous studies of e.g. second hand markets, platform users do not express critical motives for being part of online fashion sharing platforms.

So far, consumer studies have paid only piecemeal attention to consumer behavior in the sharing economy. More research is needed on how new sharing economy business models influence consumer perceptions and behaviors. From a methodological point of view, sharing platforms are often powered by new online technologies which offer interesting new avenues for gathering knowledge about the consumers. While this chapter adopted a netnographic approach for identifying consumer motivations, future studies could also consider big data analytics, tracing technologies, nudging, and online experiments as methods for deepening our understanding of consumers in the sharing economy. Last, this chapter emphasized the motives of consumers participating in fashion reselling and swapping markets. In the future, it will be relevant to explore in more detail how consumer motivations influence and are influenced by other individual, organizational, and institutional variables shaping the fast-growing sharing economy.

There are limitations to the study due to the nature of the data material and the analytical process. One potential bias concerns the sample size as this study is based on an actual sample of 363 reviews 631 reviews of mobile-enabled reselling and swapping peer-to-peer platform. Another risk concerns the translation of app review text to indication of motivation as this process is inevitably subject to interpretation. Last, there is a potential sampling bias and the risk of review manipulation by platform providers. It is likely that only highly involved platform users write reviews. It can be assumed that alternative data collection procedures might reach additional consumer groups and hence reach different conclusions. Furthermore, there is a risk that some of the studied reviews are actually fake reviews, purchased by the platform provider or generated automatically, in order to attract more consumers, with more reviews giving the impression of a mature and successful platform.

Acknowledgements The study was conducted as part of the Mistra Future Fashion project (http:// www.mistrafuturefashion.com).

References

Bardhi, F. (2003). Thrill of the hunt: Thrift shopping for pleasure. *Advances in Consumer Research, 30,* 357–376.

Bardhi, F., & Arnould, E. J. (2005). Thrift shopping: Combining utilitarian thrift and hedonic treat benefits. *Journal of Consumer Behavior, 4*(4), 223–233.

Bardhi, F., & Eckhardt, G. M. (2012). Access-based consumption: The case of car sharing. *Journal of Consumer Research, 39*(4), 881–898.

Beard, N. D. (2008). The branding of ethical fashion and the consumer: A luxury niche or mass-market reality? *Fashion Theory, 12*(4), 447–467.

Belk, R. (1988). *Possessions and self*. Wiley.

Belk, R. (2014a). You are what you can access: Sharing and collaborative consumption online. *Journal of Business Research, 67*(8), 1595–1600.

Belk, R. (2014b). Sharing versus pseudo-sharing in Web 2.0. *The Anthropologist, 18*(1), 7–23.

Belk, R. W., Sherry, J. F., & Wallendorf, M. (1988). A naturalistic inquiry into buyer and seller behavior at a swap meet. *Journal of Consumer Research, 14*(4), 449–470.

Benson, S. P. (2007). What goes' round comes' round: Secondhand clothing, furniture, and tools in working-class lives in the interwar United States. *Journal of Women's History, 19*(1), 17–31.

Birtwistle, G., & Moore, C. M. (2007). Fashion clothing—Where does it all end up? *International Journal of Retail & Distribution Management, 35*(3).

Botsman, R., & Rogers, R. (2010). *What's mine is yours: The rise of collaborative consumption*. New York: Harper Business.

Braun, V., & Clarke, V. (2006). Using thematic analysis in psychology. *Qualitative research in psychology, 3*(2), 77–101.

Codagnone, C., & Martens, B. (2016). *Scoping the sharing economy: Origins, definitions, impact and regulatory issues*. Institute for Prospective Technological Studies Digital Economy Working Paper, 1.

Crewe, L., & Gregson, N. (1998). Tales of the unexpected: Exploring car boot sales as marginal spaces of contemporary consumption. *Transactions of the Institute of British Geographers, 23*(1), 39–53.

Domina, T., & Koch, K. (1997a). The textile waste lifecycle. *Clothing and Textiles Research Journal, 15*(2), 96–102.

Domina, T., & Koch, K. (1997b). The effects of environmental attitude and fashion opinion leadership on textile recycling in the US. *Journal of Consumer Studies and Home Economics, 21,* 1–17.

Domina, T., & Koch, K. (2002). Convenience and frequency of recycling: Implications for including textiles in curbside recycling programs. *Environment and Behavior, 34*, 216–238.

Dubois, E., Schor, J., & Carfagna, L. (2014). Connected consumption: A sharing economy takes hold. *Rotman Management*, 50–55.

Flick, U. (2009). *Sozialforschung – Methoden und Anwendungen: Ein Überblick für die BA-Studiengänge*. Reinbek: Rowohlt.

Gladigau, K. (2008). Op till you drop: Youth, distinction and identity in vintage clothing.

Goldsmith, R. E., Heitmeyer, J. R., & Freiden, J. B. (1991). Social values and fashion leadership. *Clothing and Textiles Research Journal, 10*(1), 37–45.

Gregson, N., & Beale, V. (2004). Wardrobe matter: The sorting, displacement and circulation of women's clothing. *Geoforum, 35*(6), 689–700.

Gregson, N., & Crewe, L. (1994). Beyond the high street and the mall: Car boot fairs and the new geographies of consumption in the 1990s. *Area*, 261–267.

Grimm, F., & Kunze, A. (2011). Meins ist Deins [What's mine is yours] 3.0. Available at: http://www.enorm-magazin.de/leseprobe/Leseprobe_2_2011.pdf. Accessed on July 14, 2017.

Guiot, D., & Roux, D. (2010). A second-hand shoppers' motivation scale: Antecedents, consequences, and implications for retailers. *Journal of Retailing, 86*(4), 355–371.

Ha-Brookshire, J. E., & Hodges, N. N. (2009). Socially responsible consumer behavior? Exploring used clothing donation behavior. *Clothing and Textiles Research Journal, 27*(3), 179–196.

Hiller Connell, K. Y. (2010). Internal and external barriers to eco-conscious apparel acquisition. *International Journal of Consumer Studies, 34*, 279–286.

Hiller Connell, K. Y. (2011). Exploring consumers' perceptions of eco-conscious apparel acquisition behaviors. *Social Responsibility Journal, 7*(1), 61–73.

Hiller Connell, K. Y., & Sontag, S. M. (2008). *Identifying environmentally conscious apparel acquisition behaviors among eco-conscious consumers*. Paper presented at the 65th Annual Meeting of the International Textile and Apparel Association, Schaumburg, Illinois.

Hustvedt, G., & Dickson, M. A. (2009). Consumer likelihood of purchasing organic cotton apparel. Influence of attitudes and self-identity. *Journal of Fashion Marketing and Management, 13*(1), 49–65.

Iwanow, H., McEachern, M. G., & Jeffrey, A. (2005). The influence of ethical trading policies on consumer apparel purchase decisions. A focus on The Gap Inc. *International Journal of Retail & Distribution Management, 33*(5), 371–387.

Joergens, C. (2006). Ethical fashion: Myth or future trend? *Journal of Fashion Marketing and Management, 10*(3), 360–371.

Klepp, I. G. (2001). *Hvorfor går klær ut av bruk? Avhending sett i forhold til kvinners klesvaner*. Oslo: SIFO. The National Institute for Consumer Research.

Kozinets, R. V. (2002). The field behind the screen: Using netnography for marketing research in online communities. *Journal of Market Research, 39*, 61–72.

Laitala, K., & Klepp, I. G. (2011). *Environmental improvement by prolonging clothing use period*. Paper presented at the Towards Sustainability in the Textile and Fashion Industry.

Lang, C., Armstrong, C. M., & Brannon, L. A. (2013). Drivers of clothing disposal in the US: An exploration of the role of personal attributes and behaviours in frequent disposal. *International Journal of Consumer Studies, 37*(6), 706–714.

Markkula, A. (2007). *Sustainable consumption—Sustainable ways of consuming fashion*. Paper presented at the Proceedings of the Nordic Consumer Policy Research Conference, Helsinki.

McAfee, A., Dessain, V., & Sjoeman, A. (2004). *Zara: IT for fast fashion*. Cambridge: Harvard Business School Publishing.

McIlroy, S., Ali, N., Khalid, H., & Hassan, A. E. (2015). Analyzing and automatically labelling the types of user issues that are raised in mobile app reviews. *Empirical Software Engineering*, 1–40.

Morgan, L. R., & Birtwistle, G. (2009). An investigation of young fashion consumers' disposal habits. *International Journal of Consumer Studies, 33*, 190–198.

Morley, N., McGill, I., & Bartlett, C. (2009). Appendix I. Maximising reuse and recycling of UK clothing and textiles. EV0421. Technical report. Final report for Defra: Oakdene Hollins Ltd.

Netter, S. (2016). *Exploring the sharing economy*. Frederiksberg: Copenhagen Business School [Ph.D.]. (Ph.D. Series; No. 52.2016).

Netter, S. (2017). User satisfaction & dissatisfaction in the app sharing economy: An investigation into two-sided mobile fashion reselling & swapping markets. In C. E. Henninger, P. J. Alevizou, H. Goworek, & D. Ryding (Eds.), *Sustainability in fashion: A cradle to upcycle approach*. London: Springer.

Niinimäki, K. (2010). Eco-clothing, consumer identity and ideology. *Sustainable Development, 18*, 150–162.

O'Reilly, L., Rucker, M., Hughes, R., Gorang, M., & Hand, S. (1984). The relationship of psychological and situational variables to usage of a second-order marketing system. *Journal of the Academy of Marketing Science, 12*(3), 53–76.

Paden, N., & Stell, R. (2005). Consumer product redistribution. *Journal of Marketing Channels, 12*(3), 105–123.

Pedersen, E. R. G., & Netter, S. (2015). Collaborative consumption: Business model opportunities and barriers for fashion libraries. *Journal of Fashion Marketing and Management, 19*(3), 258–273.

Roux, D., & Guiot, D. (2008). Measuring second-hand shopping motives, antecedents and consequences. *Recherche et Applications en Marketing (English Edition), 23*(4), 63–91.

Schor, J. (2014). Debating the sharing economy. *Great transition initiative*.

Schor, J. B., & Fitzmaurice, C. J. (2015). Collaborating and connecting: The emergence of the sharing economy. In L. A. Reisch & J. Thøgersen (Eds.), *Handbook of research on sustainable consumption* (p. 410). Cheltenham: Edward Elgar Publishing.

Seidl, I., & Zahrnt, A. (2012). Postwachstumsgesellschaft: Verortung Innerhalb Aktueller Wachstumskritischer Diskussionen. *Ethik und Gesellschaft, 1*, 1–22.

Shaw, D., Hogg, G., Wilson, E., Shiu, E., & Hassan, L. (2006). Fashion victim: The impact of fair trade concerns on clothing choice. *Journal of Strategic Marketing, 14*(4), 427–440.

Sherry, J. F. (1990). A sociocultural analysis of a Midwestern American flea market. *Journal of Consumer Research, 17*(1), 13–30.

Shim, S. (1995). Environmentalism and consumers' clothing disposal patterns: An exploratory study. *Clothing and Textiles Research Journal, 13*(1), 38–48.

Thøgersen, J. (2010). Pro-environmental consumption. In K. M. Ekström (Ed.), *Consumer behaviour. A Nordic perspective*. Lund: Studentlitteratur AB.

Walsh, B. (2011). 10 ideas that will change the world. Time, March, 2011. Available at: http://www.time.com/time/specials/packages/article/0,28804,2059521_2059717_2059710,00.html. Accessed on July 14, 2017.

Williams, C. C. (2002). Why do people use alternative retail channels? Some case-study evidence from two English cities. *Urban Studies, 39*(10), 1897–1910.

Williams, C. C. (2003). Explaining informal and second-hand goods acquisition. *International Journal of Sociology and Social Policy, 23*(12), 95–110.

Williams, C. C., & Paddock, C. (2003). The meaning of alternative consumption practices. *Cities, 20*(5), 311–319.

Zervas, G., Proserpio, D., & Byers, J. W. (2015). The impact of the sharing economy on the hotel industry: Evidence from airbnb's entry into the texas market. In *Proceedings of the Sixteenth ACM Conference on Economics and Computation* (pp. 637–637). ACM.

Fashion as a Matter of Values. On How a Transformative Educating Process Can Initiate a Positive Change

Mélanie Sburlino

Abstract The consumer is a modern day Frakenstein, an invention of a revolutionary economic model for wealth, that has become uncontrollable, compulsively devouring goods and services. Reaching the paroxysm of «consuming» with the Fast Fashion industry, we are now wondering: «who's fault?». Since the collapse of the Rana Plaza in April 2013, Fashion activist movements, governmental associations as well as largest corporations willing to be part of the «sustainable fashion trend», have stood as the hero who would master the beast by following Nelson Mandela's credo: Education as «the most powerful weapon to change the world». But, actors for a positive change are often biased by a prosperous and vicious model, thus educating sounds more like a new soft marketing strategy or a way to moralize rather than a truthful driver. Actually, time has come to redefine the core essence of «education». A shared responsibility that involves a critical thinking learner, a challenged teacher, a meaningful message and a relevant medium, factorized by a time factor. The Y and Z generations, consumers of tomorrow, need to question themselves and initiate a transformative educating process in order to become consum'actors. Among all, through this journey, they will tend to re-balance the supply and demand law and finally become satisfied with a caring, long-lasting and circular consumption.

Keywords Consumerism · Fast fashion · Transparency · Slow fashion Sustainability · Transformative learning · Critical thinking · Consum'actor Collaborative economy social learning · Gen Z

M. Sburlino (✉)
Skema Business School, Avenue Willy Brandt, 59777 Lille, France
e-mail: melanie.sburlino@skema.edu

© Springer Nature Singapore Pte Ltd. 2019   53
S. S. Muthu (ed.), *Sustainable Fashion: Consumer Awareness and Education* ,
Textile Science and Clothing Technology,
https://doi.org/10.1007/978-981-13-1262-5_3

1 Introduction and Methodology

When it comes to thinking about education from the eyes of a powerful industry such as the Fashion industry, the first question that comes to mind is how far the teaching process can be authentic and transparent? And this is even more true considering the education towards the oxymoron «Sustainable Fashion». In fact, the frontier between educating, marketing and moralizing seems to quickly becomes the main problem to study and to solve. Regarding the trend for transparency, it appears that many brands actually stand as advocates for a positive change by making consumers knowledgeable. But in a consumerist society, this strategy appears as a poor grain of sand.

Thus, this research tried to redraw the history of consumerism to explain how we educate consumers towards Fast Fashion. A key consideration that would become the basis to understand how to unlearn ingrained values that are no more source of wellness. From a sociological perspective, the aim was then to understand why awake consumers just declared being ready to consume sustainable products but never really jump ahead. Supported by many concrete economical examples found in corporations' reports and websites, critical articles and relevant surveys, these two parts naturally led to a clear answer based on Mezirow's transformative learning theory. Therefore, the last part of this chapter mainly focused on offering a new theorized solution to transform the linear Fashion economy into a circular and collaborative Sustainable Fashion economy.

2 Context: On Buying Binge

2.1 The First Consumer Revolution

The ability to easily buy needless stuff is quite recent regarding the definition of «consumerism» given by *The Cambridge Dictionary*: «a state of **an advanced industrial society** in which a lot of goods are bought and sold». Thus, the early stages of the consumerist phenomenon can be dated back to only three decades ago, in the late 18th, when the first Industrial Revolution deeply reshaped our economic and social logic.

In fact, we could even observed a genesis of such a consumerist culture at the turn of the century (Booth 2016). For most of history, people have owned more or less nothing. They were satisfied with the bare necessities. However, in the early 18th, economies of the North–Western Europe started to expand as aristocrats were willing to get rid off the mass (Coquery 1998). Spending money in fashion and luxury goods was a categorical imperative to show off their social status. These highly-demanding consumers have enriched the merchant bourgeoisie who has therefore gained a higher purchasing power to also afford small luxuries—*meaning inessential, desirable items often expensive or scarce, that they were not able to buy before*. This growing middle-class embracing new ideas about luxury consumption was particularly attracted by the eccentricities of aristocratic families. A nascent desire mainly fostered by the remote ancestors of the Fashion press. For instance, in Great Britain, *The Gallery of Fashion* and *The Ladies Magazine* were the two main references for everyone who wanted to be aware of the latest trends for clothes and hair, now altered every year (Booth 2016).

Thus, the growing importance of fashion, as an arbiter for purchasing rather than necessity, added to the development of both industrial production, trade and consumption propelled what we called **the first consumer revolution**. A virtuous economic cycle nurtured by an increasing variety of goods that was seen as a bonanza. Indeed, in 1723, Bernard Mandeville, a London physician, explained in his essay «*The Fable of the Bees*», that what makes countries rich is simply shopping for pleasure. An economic theory that could only work if a change in consuming patterns occurs as happiness and materialism was not so correlated at that time. Thus, governments and companies had to educate people to become demanding consumers of non vital stuff (Ebeling 2016). That is why the emergence of many of the institutions that sustain and promote mass consumption can be dated to the late 18th century. Department stores appeared, attractive packaging were developed, historical brand names were launched and advertising started to rival with education and religion in shaping people's values and aspirations.

This theory has been vehemently criticized by The Christian Church and conservatives, such as the philosopher Jean-Jacques Rousseau who called for a return to a simpler way of life, far from those vain and materialistic behaviors. He even proposed to impose taxes on luxury goods in order to refocus people's minds on non-material values (Booth 2016). From those times is born the conflict between wealth/consumption and poverty/virtuous restraint. Then, the Enlightenment philosophical movement combined with the Great Famine that affected the European countries in 1845 have consequently slowed down this frenzy, which only re-upsurged with the Second Industrial Revolution in 1870 and then fluctuated with major historical events.

2.2   The Golden Age of Consumerism

1930s. United States of America. The consumerist society began to revive from the Great Depression, later defined as the worst economic downturn in the history of the industrialized world. The stock market crash that occurred in October 1929 led to the creation of a vicious circle: the dramatic drop of consumer spending and lack of investments … causing the decline in industrial output … causing a vague of unemployment … causing the loss of purchasing power, and so on. The solution to break this infernal cycle? **Consumerism**.

At that time, the now so polemical concept of «consumer society» was seen as *a necessity* to relaunch the economy. In fact, politicians, economists and businessmen came back to the idea developed by Mandeville: «shopping for pleasure», a new need for people who could afford it. But the schema was quite different this time as most of those people already own basic wares. So, why would they spend money for stuff they already have and moreover, that were working pretty well? Two strategies answer this question, and above all, led to the creation of our current unstoppable economic model.

Firstly, manufacturers attracted wealthy potential buyers by «educating» them through advertisements. The 60s were the apogee of the advertising industry. Companies and advertising agencies were fighting for naming rights of public facilities. The challenge? Be everywhere to inscribe their brand in people's minds. From television to newspapers and magazines, contents were flocking to inform people of new products, new functions, new design. Advertisers were delivering a crystal clear refrain: «New is in, old is out». Then started the run for the latest trends. In fact, more than just delivering factual data, media hype was promoting a dreamed vision of the good way of life, combining the idea of happiness with consumption, associating success and social recognition to the ability of buying things, a powerful message regarding the past years of recession. Since that time, consumerism has been intimately associated to the belief that wellbeing depends on the purchase of material goods. The United States was the paroxysm of a hyper-consumerist society nurtured with a relentless flow of advertisements urging to buy ever more. The main target were the easiest one to shape: young children. Educate to the living source. Indeed, according to a 2007 survey realized by the American Federal Trade Commission, children aged 2–11 in the late 1970 spent over a week of their lives every year watching ads. How did brand educate children to become perfect little consumers? Lucy Hughes, director of communication strategy at Initiative Media, developed the theory of the «Nag Factor», a simple concept explaining the power of shaping right nagging behavior to make parents buy what children integrated they should desire. Therefore, this generation of children grew up with this taken-for-granted mentality, in a new era of consumer sovereignty over capitalism where buying is power. From supposedly «passive buyers» to «full-rights consumers», a true culture of buying binge has emerged, «consumption as a natural human behavior» taught from early childhood by advertisers/companies, and indirectly by politicians and economists.

Secondly, the challenge was to keep a constant demand. Buying according to trends logically implies that the good you acquired would have a lifespan determined by the time for markets to define a new trend. But if this assumption would appear as obvious for our current society, this behavior was not yet automatic for the previous generation. A risky stake unthinkable in times of Great Depression. So manufacturers thought about an additional driver to foster people to replace their belongings. They implemented what we call today «planed obsolescence». By using poor quality raw materials, deliberately designing objects with failures or even avoiding to provide repair manuals, manufacturers progressively reduced the lifespan of what we buy, in order to shorten the repurchase cycle. By the 1950s, it has become the major paradigm in mass production. Indeed, looking back to the pre Industrial Revolution era, people's possessions were limited, mainly handmade by the household members themselves or artisans. People's clothing, for example, were used, repaired if needed, for decades. Classical items such as winter coats were expected to last a lifetime and more, they were even passed from one generation to the next while today, a winter coat often last just one season (Goodwin 2008).

Thus, fostered by the call of the new, people started to buy more. More stuff designed to not last, that they needed to replace earlier. This phenomenon has shaped the main common definition of «consumerism»: the continual expansion of one's wants and needs for goods and services. And today, we can observe that the idea that something broken, or just outdated should be throwaway and replace has slowly but surely been ingrained in our mind. It has become a mechanism. Economists, politicians, manufacturers, advertisers have created a kind of modern Frankenstein. Consumers were a revolutionary invention to revitalize the economy but have become uncontrollable, reaching the paroxysm of a term: burning, destroying, devouring, wasting. Vance Packard, a forward-thinker writer, warned about the negative aspect of such tornado explained in his book *The Waste Makers*, published in 1960, that consumerism can just be defined as «excessive materialism and waste». But this early warning was a whisper drowned out by a recovered prosperity.

2.3 Fast Fashion, a Consumerist Model at Its Paroxysm

The whole story has started with Fashion. Indeed, it is the British textile industry that initiated the development of large-scale industrialization in 1734, when John Kay invented the flying shuttle (Goodwin 2008). From 1760 to 1830, the annual amount of cotton used in that industry rose from 3 million pounds to more than 360 million pounds. From now on, the challenge was to create the demand in order to consume such output. As in 1800, England only counted 7.7 million inhabitants, of which only a small minority could buy under color of Fashion, they began to export on bigger markets such as India. Thus, the question is: how Fashion has become such a social phenomenon in Occidental countries, a consumption driver deeply ingrained in every single individual's life?

According to a McKinsey report, the Fashion industry is now worth $2.5 trillion and is standing as the world's seventh-largest economy if ranked alongside individual countries' GDP. Looking at the brief history of Fashion, it appears that its development as an industry has emerged in the early 20th century with first, the genesis of great figures in the luxury sector, and later, the creation of super powerful brands. By adding an industrial dimension to clothing creations, couturiers such as Chanel, Poiret or Schiaparelli, managed to deeply changed the economic model of Fashion, moving from a world ruled by nameless craftsmen to a world dominated by big names renowned all around the world.

Nevertheless, the clothing production approach was still anchored in this conception of Fashion as exclusive. A vision supported by the thinking of the Silent Generation (from mid-1920s to the early-to-mid 1940s) whose life motto was «Repair, Reuse, Make do, and don't waste anything».

The genuine emergence of the Fashion industry, as defined by *The Oxford English Dictionary*, really took off after the Great Depression and the Second World

War while consumption norms started to evolve. Mass production implemented to relaunch the economy led to mass media to create demand. Then, the mass consumption phenomenon started, to stop at nothing. This rupture between promoting a style for an elite and producing and marketing new styles of goods for a mass market is directly linked to the end of times of uncertainty and deprivation, a way to celebrate a rediscovered freedom through consumerism. Fashion has always been frivolous, but only for those who had the time and the money to be so. Now, everyone could enjoy the Fashion playground as the purchasing power has been relaunched. It was a liberator movement. A revolution.

And the raiding patrol of this revolution was driven by Marketing. Indeed, everything Fashion does is predicated toward teaching consumers «to eschew the old and embraces the new» (Friedman 2013). A thought that totally fitted the philosophy of the advertisers of the 60s. Ad campaigns were massively promoting trends to attract consumers while branding, a key component of Marketing, aimed at retaining a clientele by creating connections between customers and products. This unique relationships developed the idea that «*I buy therefore I am*». Consuming became a matter of feelings, what totally disturbed the original Maslow's hierarchy of needs. While buying few clothes was seen as a psychological need, from the 1950s, Fashion began to infiltrate all the stages of the pyramid. Just ask yourself: why do I buy the things I do? Because it is vital? Or because consuming Fashion goods has become a source of well-being (safety needs), a way to feel socially accepted, to belong to a «community» (social belonging), an imperative to be valued by others, to be noticed and admired, maybe also a way to boost your ego and to accept yourself (esteem), a mean to create a better version of yourself (self-actualization)?

The success story of Erling Persson, founder of the giant Swedish H&M, perfectly illustrates how far companies have been involved in the change of consuming patterns (Bowman 2016). This forward-thinker got the turning point while traveling through the USA, which was actually recovering from a tough economic time. Amazed by the dynamic and trendy Big Apple, he came back in Sweden in 1947 with the idea to create something huge, a new institution for Fashion, a pioneering way to sell clothes. His main goal was to democratize Fashion, to make it easily affordable. To make it fun by offering an infinity of choice. He imported the way of life of New Yorkers, who turned shopping into a proper leisure. A vision depicted by Blake Edwards in 1961 through his romantic comedy film, *Breakfast at Tiffany's*, starring the elegant Audrey Hepburn. Persson was a true child of the century, he literally applied the economic model of mass production that was slowly but surely shaping the new society's behaviors. The fast development of the brand can be observed with the opening of the second Hennes store in Stockholm in 1952. Followed by the opening of the first store in Norway, in 1964. Today, H&M is located in 69 countries and owns more than 4700 stores. They strongly participated in the spread of a worldwide mass-culture through their distribution plan. The extension of the brand's activity also showed the need to produce more and more. Thus, the company went from producing womenswear to selling menswear and children's clothing. In 1977, a new make-up line has been launched and in 2000s, H&M Home was born. H&M exploded every opportunity, irrigated the world with every possible channel. To produce such a quantity of goods, they built a massive network of factories and pointed out the most populated areas to benefit from the strongest workforce. Thus, it is not surprising to notice that H&M worked with 449 Chinese manufacturing factories and 166 Indian ones—*the two biggest countries regarding demographical figures*, over 1681 manufacturing factories in total. In other words, China represents 27% of the manufacturing force of H&M and India 10%. But mass production is not the only factor that explains the success of Erling Persson's venture. In fact, it was all about producing **faster**. So, the master draw from the principle of everything that we actually explained before: «planned obsolescence». Or maybe, the word «trends» would be more appropriate when talking about Fashion. He used the notion to the core, constantly releasing cheap collections in record time. That is why we are now used to be fed with weekly to daily new arrivals. The system described through the history of H&M is nothing else than what we call today: Fast Fashion. An accurate expression to appoint the quick renewal of fashion collections (Martin 2017). And when Zara joined the movement in 1974, the Fashion industry assisted to a true state of siege by the Fast Fashion. In fact, regarding the stock market, H&M, Zara and Mango, the three leaders of the Fast Fashion industry, are now equal worth to high-end brands. Luxury houses even look at the mainstream fashion to retarget their strategies. If Fashion was before considered as being the onerous scarcity, it is now seen as the perpetual novelty. From two collection a year, we can now find happiness in six to eight collections a year (Martin 2017). Zara is even producing more than 12,000 different models of clothes per year! And consumers, who turned the new into a true religion, set the pace by buying binge.

However, we could have expect the phenomenon to slow down regarding the past economic fluctuations. But, it seems that people acquired the taste of consumerism. The demand has taken over from the offer. Thus, when the financial crisis of 2008 burst—*a global downturn described by Ben Bernanke, former Fed chief, as worst than the 1929 crisis*, the Fast Fashion industry managed to minimize the impact thanks to a strong low price positioning. To «mass» and «fast» has been added a new characteristic: «cheap». A loss in value that has fostered a throwaway mentality (Vince 2012). Thus, through the 20th and early 21st century, we developed a linear economic model based on the use of massive resources to produce quantity and infinite choice of clothes and answer an endless demand for new. A stream of fashion goods that has inevitably lost value, both in terms of price than in terms of affect. Fashion has therefore become disposable.

But, to what extent can it last? How can such a bulimic economic model not explode? By continuously moving up a gear, do we not race headlong into a concrete wall?

3   Awareness: An Out-of-Fashion Model

3.1   Why Our Current Economic Model Is Intending to Be Consumed

The Fashion industry is a veritable chameleon. Perpetual renewing is even the core essence of Fashion. As a universal social phenomenon and a global economic driver, this industry has always managed to adapt itself to the most significant changes. Thus, the evolution of the Fashion industry is reflecting the answer to the needs and wants of a society at a particular moment of the history. In the 19th, its emergence was directly correlated to the coming of a core idea: individual as the highest value (Monneyron 2006). The cycle was then updated every twenty years, going from the Roaring Twenties, marked by the quest for a new form of liberation, breaking the rules of formal fashion to adopt comfortable clothes, to the height of Haute Couture in 1940, then came the internationalization of Fashion in the 1960s, the birth of ready-to-wear in the 80s and finally, the reign of Fast Fashion established in 2000. So, what is gonna happen next? Will we continue to follow the diktats of Fast Fashion a further twenty years?

Regarding the unsustainability of such linear model, the end of the Fast Fashion's reign is calling. A vindictive reign with 80 billion clothes consumed every year, 20 pieces of clothing per person each year, 400% more than what we consumed two decades ago. And the change in consuming patterns is even more recent. Indeed, according to Mckinsey & Company, the average consumer bought 60% more clothing in 2014 than in 2000, while he kept each garment half as long (Drew and Yehounme 2017). We made consumption our way of life, the act of buying has even become a ritual as we are constantly seeking for a kind of spiritual satisfaction, an «ego satisfaction». The major need of our today's society is to consume things, to burn up, to worn out, to replace, to discard at an ever increasing rate (Goodwin 2008). How have we managed to reach such a point? By implementing innovative production and distribution models to offer clothing cheaply and quickly. Thus, consumers easily come to the conclusion that the white cotton t-shirt they bought two weeks ago is now out-of-fashion and worn due to the poor fabric quality. Then, as it was cheap, why not just throwing it to buy the latest version? Maybe because, the t-shirt you bought at Zara for only $9.90 is worth more than the price you actually paid.

In 2015, *The True Cost*, a poignant documentary directed by Andrew Morgan, highlighted the pungent issue of Fast Fashion and the hidden part of our appetite for cheap prices and novelties. Indeed, the Fashion industry is the second dirtiest industry in the world, highly responsible for water stress, water pollution and climate change. Just a simple fact: cotton represents 33% of all fibers used to produce clothes but one single cotton t-shirt is worth 2,700 L of water, enough to satisfy the physical needs of one person for two years and a half! And to respond to consumers' bulimia, we developed cotton farming activities. Thus, today, 90% of the cotton used by the textile industry is genetically modified, using vast amount of water and chemicals. Cotton farming represents 18% of worldwide pesticide, 25% of insecticide despite using only 3% of the world's arable land. Fabric dyeing is also involved in water stress—*5 trillion liters of water are used each year to color our clothes*, and pollution—*being the second largest polluter of clean water globally*. These are just few examples to illustrate the role of Fashion in climate change. But above all, the darkest side of Fast Fashion is maybe the issue of textile waste. The use of such an amount of resources appears pointless considering that consumers throw away an average of 30 kg per person annually. Thus, the average lifetime of our clothes declined from several generation to three years (LeBlanc 2017). And the life expectancy of the workers producing Fashion followed this path.

24 April 2013, Rana Plaza collapsed. Killing 1,134 people. Leaving 2,500 injured (Westerman 2017). This over crowded building located in Dhaka, Bangladesh, with too many floors, too much heavy equipments to produce ever more, faster, cheaper, housed five garment factories working for major Fast Fashion retailers such as H&M, Gap, Primark, Benetton, Walmart and Mango. Indeed, 90% of the clothing for Western consumers is produced in South-East Asia, especially in China, Bangladesh, Vietnam, India and Cambodia. Why? Because it represents a strategic choice to offer low cost fashion garments due to the cheap labour force. In fact, the wages regulations are much more flexible in this area. Thus, today while Bangladesh—*considered as the world's workshop,* weights $28 billion in the garment industry, the average basic monthly pay of a Bangladeshi worker is only $80—*no more than 1–2% of the price of an article of clothing*! According to a survey of Bangladeshi factories, the average estimate of what workers consider enough to live on and support their families is $170 a month (Parry 2016). That is modern slavery. Beside the below-subsistence-level wages, the Fast Fashion industry is responsible for unethical working conditions in the production plans: safety standards are often violated, workers suffer malnutrition while facing ever more work pressure and they have no freedom of association. Thus, if the «*luckiest*» generally start working at the age of 14, dedicating their whole youth to multinational retailers willing to feed our appetite for trends, after they reach 40, they have to retire because they become too old to fulfill the production targets.

Unfortunately, we kind of *needed* the Rana drama to reveal the true cost of our cheap clothes, to raise the question about transparency and to draw awareness of inhumane working conditions. Time has come for a Fashion Revolution.

3.2   The Call for a Positive Change

Oscar Wilde has always been recognized for being a forward-thinker, and he kind of perfectly described, more than one decade ago, in one short sentence, the main problem we are now facing, enrolled in the Fast Fashion whirlwind: «Nowadays people know the price of everything and the value of nothing».

After the collapse of the Rana Plaza, initiatives have emerged to measure and highlight this hidden value. Government agencies, activist associations, scholars, journalists, the UN gathered to form one strong single voice urging to consume responsibly, to stop taking everything for granted and think about the whole pre and post process of making cheap clothes. But, despite the brutal awakening, those too environmentalist and moralist speeches did not take root into consumers' mind. In fact, we can notice that the craze for cheap and fast has not slowed down. According to Inditex's official annual reports—*the Spanish group includes 8 brands: Zara, Pull&Bear, Massimo Dutti, Bershka, Stradivarius, Oysho, Zara Home and Uterque*: sales would have increased from 16.724 billions euros in 2013 to 23.311 billions euros in 2016 (Inditex 2017). If the first part enabled to understand how the Gen X (1960–1980) used marketing—*especially advertising,* to ingrain consumerist patterns in our society, now the question is raised: why the new generation, which built itself in opposition to their parents, carries buying binge on while being informed of the consequences of such a behavior? Just looking at Millennials' favorite brand, Apple, we get the point of this demanding but not regarding generation: they want yet accessible premium goods, yet approachable savvy stuff, innovative products the brand will obsolete itself, and after all items combining commerce and conscious (Hess 2011). According to this last assumption, Sustainable Fashion should totally meet the expectancies of Millennials. They even consider it as cool (Pringle 2018). Nevertheless, studies have showed that despite the demand for socially responsible fashion, Millennials always rank the «price» and «ease of purchase» factors much higher than any sustainable arguments. Thus, according to a survey published by Deloitte in 2017, only 2.6% of millennial luxury shoppers across China, the US, the UK and Italy consume driven by brand's ethical standing (Pringle 2018). The main problem is that this «IWWIWWIWI generation», aka «I Want What I Want When I Want It» as called by the Financial Times Fashion journalist Vanessa Friedman, feel powerless in front of the guilt of Fast Fashion as this economic model has become the norm. Of course, as digital natives, they have an easy access to information to measure the responsibility of the biggest apparel leaders and to find out alternatives but the fact is that their taken for granted and demanding mentality pushes them to ask the brands to make the positive change. They are not ready to initiate it by themselves.

Maybe because activists focused more on pointing the Fast Fashion industry out, instead of also including consumers. Indeed, since 2013, millions of articles have been published to list the brands responsible for unethical behaviors. Greenpeace among other NGOs has multiplied protestations and campaigns such as the Detox campaign exposing the direct links between clothing brands, their suppliers and toxic water pollution (Perry 2018). In 2017, the French National Assembly even voted the Duty of Care Act (*Law 2017-399*) in order to impose the largest corporations to adopt a vigilance plan to forestall human rights abuses and environmental damage. This vigilance plan is applicable on the whole supplier chain, on subsidiaries as well as on subcontractors located all over the world. By doing so, all these organizations have focused on raising awareness, probably the first step in consumer education. They push «the bigs» to lead the way, they foster consumers to hold the Fast Fashion leaders to account.

3.3 Green Is the New Black

As the calls for a positive change were mainly addressed to Fast Fashion retailers, a turn in brands' communication strategies has been observed over the past five years. An increasing number of global corporations answered to activists' claims for consumers' right to know, and disclosed information about their first-tier supply chain. According to the Fashion Revolution association, the number of brands listed on the Fashion Transparency Index has increased from 5 in 2016 to 32 in 2018. Nike Inc., H&M Group, Levi Strauss & Co., Abercrombie & Fitch, Asos, Benetton, … are among the first to have been involved in the Transparency Pledge campaign (Ditty 2018). This *trend* for more transparency is crucial to help consumers to better understand the current economic model that is now involving numbers of intermediaries. By publishing the traceability of the clothes we buy, we can realize that behind the apparent easy access to Fast Fashion, a quite complex process exists. This additional knowledge is also a tool to pique buyers' curiosity on the working conditions and income threshold applied by those partners. Willing to clean up their acts, brands also highlight their new sustainable and social responsible strategies both on their website and annual reports. For example, in Zara's 2016 Annual Report, we can see that the «This Year's Highlights» part is particularly focusing on communicating a responsible image. Thus, in June, «The 2016 Newsweek Green Ranking rated Inditex as one of the world's most sustainable listed companies», in July, «Inditex unveiled its 2016–2020 Environmental Plan which delves further into the Group's unfolding circular economy model, tackling every link in its productive model» and «Greenpeace ranked Inditex as an Avant-Garde performer in its 2016 Detox Catwalk for its use of sustainable chemicals», and finally in January 2017, «Inditex took the gold metal in the Dow Jones Sustainability Yearbook 2017 for its sustainability performance» and «The Group reinforced its agreement with water.org for bringing drinking and toilets to vulnerable communities». Then, the brand develops its «Sustainable Strategy» and «Priorities» for people, recycling, efficient use of resources and community welfare during almost one-third of the report.

In addition, some companies went further by transforming their business model to surf on the green wave. Many Fast Fashion groups such as H&M, Zara and Asos launched «eco collections». In fact, H&M is leading the way with the H&M Conscious Collection. In February 2018, they even released an Exclusive Conscious Collection made of sustainable fabrics such as organic cotton, linen, silk, tencel, recycled polyester, recycled silver and Econyl, a fiber created from fishing nets and waste nylon. On its part, Zara has introduced «Join Life», a collection «that respects the environment» and reduce their environmental impact. And Asos, started an ethical and solidary Made in Kenya collection by collaborating with the Kenyan factory, Soko Kenya committed for a sustainable, environmentally-friendly and local production. Through fresh new and nature-inspired campaigns, the giants of Fast Fashion show their new credo for sustainability and stand as green gurus leading consumers on the straight path. Not only offering alternative way of dressing, they also *educate* consumers to «close the loop» as the 2015 H&M campaign claimed. Indeed, more than 33 Fashion companies joined in a pledge to increase their clothing recycling by 2020 (Drew and Yehounme 2017). That is why it is now common to find point of recycling abandoned in the corner of Fashion stores.

But disruptive concepts have also emerged to offer new alternatives for fully responsible consumers. We are all aware of Patagonia, the pioneer of quality clothing with organic, recycled and up-cycled fibers; Veja, this French sneakers brand which uses fair-trade rubber and organic cotton; Reformation, a Los Angeles-based clothing brand that produces clothing with far less water and emissions than typical clothing companies; Everlane, which has for mission to offer a fair and ethical price and committees to publish the detailed cost of what it sells; or even the higher-end brand, Edun, who brings sustainability on the catwalks. Those brands are the advocates of a new movement called the Slow Fashion, an expression defined in 2007 by Kate Fletcher in The Ecologist.

So, we have new sustainable Fashion stores, new sustainable Fashion collections, new conscious campaigns, new companies for more choices, …? Everything new, reminding the speech of the 30s that launched the consumerist revolution. Now, let's think a second … Can Fashion be sustainable? The New York Times fashion critic Vanessa Friedman would answer: «No way». During the 2016 Copenhagen Fashion Summit, she argued: «Sustainable Fashion is an oxymoron as fashion is about producing and marketing new styles of goods while sustainable is about maintaining at a certain rate or level […] On the one hand, we have the pressure to be new; on the other, the imperative to maintain» (Chua 2014). Actually, Inditex&Cie expertly play on words, promoting sustainable actions while continuing to release amount of new clothes every week. Sustainable has become the most effective marketing word to make sells, it has become a trend. A competitive advantage used by everyone and anyone. What was supposed to revolutionize consuming patterns by informing, offering alternatives and educating consumers to adopt a responsible and circular behavior, is in fact creating confusion, diverting consumers with bright greens, and striking statements for change. That is no more than *greenwashing*, the promotion of environmental and social initiatives without the implementation of business practices that would really minimize the negative impacts of fast fashion. More than misleading consumers, it breeds an atmosphere of mistrust amplifies the difficulty to make a wise purchasing decision as people are discouraged to check every information published, every real meaning and impact of the word used by marketers. And when the bravest try to go deeper into companies' manifestos and arguments, they always find a closed door at one point as a huge lack of transparency is still running the game.

Finally, what would be the successful recipe to truly and fully educate consumers towards more Sustainable Fashion? When NGOs, governmental associations, individual activists moralize, we do not feel concerned, when large corporations market, we feel foolish … Deadend? No, in fact a starting place to answer will be found in the core essence of what education is, a definition too often forgotten.

4 Sustainable Fashion: A Shared Responsibility

4.1 Towards a Transformative Educating Process

The Cambridge Dictionary defines «education» as «the process of teaching or learning». However, regarding the previous research developed all along this chapter, education can only become a key driver for a positive change if a small but relevant shade is added to this primer definition. The purpose is not to teach or learn, but to teach **and** learn. Indeed, to become the «powerful weapon» Nelson Mandela was advocating for, education needs to be seen as a circular process where the teacher and the learner work collaboratively, respond to each other to transform the delivered knowledge into actions. Thus, education could be defined as an unalterable equation including (1) a collaborative work between a learner and a teacher, (2) a meaningful message, (3) a relevant medium, factorized by (4) the time factor. It is all about building a shared, meaningful and durable commitment.

An assumption directly linked to Jack Mezirow's **transformative learning theory**. This American sociologist and author of *Learning as Transformation: Critical Perspectives on a Theory in Progress*, first published in 1978, was actually studying the learning methods developed by adults re-entering higher education, methods necessarily different from those applied to educate children as youths' blank brain is generally more workable. In fact, when individuals are facing a disruptive view of the world—*the new philosophy of Slow Fashion vs. Fast Fashion based on consumerist behaviors ingrained since 1930 in people's meaning schemes*—drawn from a personal experience or another person—*a teacher, a journalist, a company, an activist, etc.*, they cannot just acquire knowledge, they need to make an additional effort by firstly reviewing their frames of reference (beliefs, attitudes, emotional reactions) in order to turn this knowledge into a driver for change. Mezirow called this disruptive factor which initiates a transformative learning process, a «disorienting dilemma», and defined it as an experience that do not fit into a person's current beliefs about the world. People having experienced such a disorienting dilemma will then need to reconsider their beliefs through a «critical reflection» in order to redefine their inner values and go from knowledgeable learner to active learner (Howie and Bagnall 2013). This critical reflection process aims to generate three core transformations: firstly, a psychological change to understand the self, secondly, a convictional change to revise the belief system and finally, a behavioral change to concretely transform the lifestyle. Mezirow's actually summarized the main distinction between a basic learning process and a transformative learning process in a clear paragraph exposed in the following quote: «A defining condition of being human is that we have to understand the meaning of our experience. For some, any uncritically assimilated explanation by an authority figure will suffice. But **in contemporary societies we must learn to make our own interpretations rather than act on the purposes, beliefs, judgements, and feelings of others**. Facilitating such understandings is the cardinal goal of adult education. Transformative learning develops autonomous thinking».

After having exposed the theoretical facts of transformative learning, it is necessary to define the key elements needed for moving from principles to practice and find out relevant solutions to the main question this chapter is dealing with: «How to educate consumers towards Sustainable Fashion?». As previously explained, collaboration is at the heart of a transformative education process. Thus, the role of educators is not negligible. They have to assist learners through the whole process which purpose is to make them aware and able to question their own frame of reference. In order to do so, teachers have to foster a learning environment by providing opportunities for critical thinking, opportunities to relate to others going through the same transformative process and opportunities to act on new perspectives (Esthermsmth 2017). This environment must be trustworthy and careful. If educators do not believe in what they are advocating and do not demonstrate a willingness to learn and change, learners will be quickly discouraged. Thus, achieving transparency, finding a common language and getting feedback is essential to ensure an efficient education. It is only on this basis that learners will construct knowledge about themselves, others, social norms and continue to animate the learning process. Finally, across this journey, consumers will educate teachers in return by questioning them, challenging them and experiencing critical discussions.

There is one particular organization which aims to conduct a transformative learning process to improve the current Fashion industry: the nonprofit global movement Fashion Revolution, launched right after the Rana Plaza disaster in 2013 by two former designers, Carry Somers and Orsola de Castro. Its members' mission is to foster greater transparency, sustainability and ethics in the Fashion industry. By being truly committed and driven by authentic beliefs, they applied the most important factor for creating a trustworthy learning environment: «demonstrate a willingness to learn and change». Moreover, disconnected from any government institution or corporate groups, they stand as trustworthy advocates for a positive change. In order to radically change the way our clothes are sources, produced, distributed, purchased and used, they aim to assist citizens, brands, retailers, students, producers, educators in each step of their transformative approach. Present in over 100 countries around the world, they encourage practice in redefining the issues linked to the Fast Fashion industry from different perspectives. For example, they raise awareness by regularly posting articles on their blog, by realizing podcasts, by publishing reports—*Fashion Transparency Index,* and editing a bi-annual free-access Fanzine—*the first one «Money Fashion Power» tackled the garment workers' conditions introducing a yearlong research project about the lives and wages of hundreds of garment workers in Cambodia, Bangladesh and India while the second edition «Loved Clothes Last» is about the issues of waste and mass-consumption in the Fashion industry.* But the Fashion Revolution association does even more than just informing through snackable contents, they foster critical thinking by introducing new ideas and engaging people to participate in a strong community of activists united in a shared experience—*123k followers on Instagram.* Thus, through their website, social media platforms, campaigns and events, they call for action and offer the opportunity to engage as this year Fashion Revolution Week's slogan expresses: «Be curious. Find out. Do something». They managed to relate people to others by implementing methods such as consciousness raising, life stories and participation in social action. For example, during the Fashion Revolution Weeks, they encourage people to ask brands *#whomademyclothes* through Twitter, Instagram, email for greater transparency. They also launched several challenges such as the *#haulternative challenge* to promote a way of refreshing a wardrobe without buying new clothes or the *#lovestory challenge* calling people to fall back in love with the clothes they already own and involving them to share their story and inspire others. This global movement has managed to efficiently communicate thanks to the massive use of social media and a smart adaptation to Millennials and Gen Z's language and taste for visual and attractive contents (Saner 2017). Nevertheless, if this global organization offer all the opportunities needed for a transformative learning environment encouraging consumers to review their inner values and act on new perspectives, the true transformation has not yet taken place. Educators having built a strong basis, learners now need to actively take steps that acknowledge these new frame of reference.

4.2　From Consumers to Consum'actors

During a 2016 TEDxSydney Talk, trying to answer the question «How to engage with ethical fashion?», the designer and activist, Clara Vuletich, argued that «Sustainable Fashion was all about values». She pointed out the fact that we have cut the process short by pointing the finger at a brand for not having increased a transparent supply chain or ethical corporate social structure, at the governments for not having increased the minimum wage threshold or not having implement a landfill tax, or even at consumers for buying amount of clothes and then just throwing them. But what we have not done yet is thinking to what extent our own responsibility, as an individual, is impacting the behavior of the whole community. She highlighted the need to, first and foremost, redefine our own personal values, our own frame of reference. This is the only way to feel empower and finally, positively change the Fashion industry. Actually, she described the second step of Mezirow's learning process. Once an adequate learning environment is built, time has come for learners to act. And the first action they can do is to question their current meaning schemes.

　　Thus, consumers would firstly need to re-frame their consumerist beliefs and behaviors, asking themselves «How did we come to buy more than the bare necessities? Is it my own choice? Do I really want/need to buy this? Or is it someone else who want me to buy this? Why?». By questioning themselves and others' assumptions, they will manage to reach their real motivations and become more responsible consumers. They will learn to unlearn what has been taught for more than a century. Indeed, as seen in the first part of this chapter, our entire economic system let us think that acquiring stuffs is the key of happiness. But has more consumption really led to make people happier? No, it has just led to a disease of affluenza (Goodwin 2008). In fact, a US survey showed that in 1957, 35% of respondents said they were very happy. Between 1957 and 1998, the average purchasing power of the average American citizen almost doubled. Nevertheless, in 1998, the proportion of people saying they were very happy decreased a little to 32%. The act of buying is not providing durable satisfaction as there is always someone with more than we have, there is always something new better than what we bought. Our education system focuses on how to give people the skills and knowledge to get a good and well-paid job in order to buy more stuff and become happier. We sacrifice our time for steady income that we will spend by buying binge that would not satisfy needs we did not question. Needs that do not really correspond to our own meaning schemes, a mirage created by the consumering whirlwind promoted by advertising. Being aware of this nonsense fact and willing to highly increase our sense of well-being—*even before thinking of the planet and others' well-being*, we can start a learning process to become more knowledgeable and feel empowered about how we are going to spend our money. Spending successfully means to understand ourselves pretty well to only acquire the material things which will contribute to increase wellness (Booth 2016). We need to register our emotions when experiencing a purchase to adjust our spending patterns. And then, to come off trends, to stop buying according to what others are buying, to stop agreeing corporations' marketed vision of happiness. At a time marked by individualism and customization, standing for our own sources of happiness should be easier. The challenge is to train on how to focus on the sources of our real satisfactions—*the psychological change*, and then, on how to hold on to these even when they lack external validation—*the convictional change*. Then, we might naturally find ourselves slightly re-defining our priorities, discovering that many of what we own do not bring us any pleasure while a few really do make us happy. Therefore, the purpose is to focus on the appreciation and expenditure of these few things. During the 2014 Copenhagen Fashion Summit, Vanessa Friedman well expressed the pleasure of building a wisely chosen wardrobe. She referenced her grandmother's passion for clothes and said: «Once upon a time, my grandmother saved and saved to buy a nice leather handbag, and once she had it, she had it for decades. Her fur coat? Same story. Her cashmere sweaters … […] same. She knew how to wash her garments—by hand usually—and how to hand them, and how to store them, be it for next season, or the next generation. What she had—what she built—was a sustainable wardrobe. […] **It is about emphasizing the value proposition** inherent in each item you buy and consciously selecting it, maybe because it has an ethically conscious aspect you appreciate, and you bothered to research the supply chain […], or maybe because you know the amount of handwork that has gone into it and you are amazed by the artistry or even know the artisans […], or maybe because you know that cashmere came from happy prancing goats running free on the steppes of Mongolia—whatever. The point is that the decision about what constitutes value in yours and you need to make it. And that implies some level of investment over—and in—time.» (Friedman 2013).

The behavioral transformation will then occur when consumers will reflect their own value system in their act of consumption. By doing so, they will become consum'actors, meaning active consumers willing to make responsible purchase acts based on knowledge and critical thinking (Rahmil 2014). Consumption will become a meaningful act defining the way in which we wish to positively change the world.

Moreover, adding to this sense of responsibility, these new consumers will develop a sense of power over the leaders of the Fashion industry.

4.3 Collaborate to Close the Loop

One of the major point part of the transformative learning theory is ongoing dialogue. Dialogue between the learner and others—*as part of a community of learners sharing the same transformative experience*, but also dialogue between the learner and his/her teacher. Thus, the trigger which will have the strength to turn the whole $2.5 trillion Fashion industry is not the consumer alone, but the union between educated consumers and the challenged industry itself. An assumption based on the historic law of supply and demand developed by Adam Smith in 1973. Much of his theory was actually arguing that everything we value about the functioning of an economy could be found in people's demand. Thus, once again, a lot of responsibilities are put on consumers' shoulders, an idea that can be easily shade. While most of economists tend to argue for consumer sovereignty, raising consumers' satisfaction as the ultimate goal to reach, nowadays, some others would better soften this concept and claim that consumers' behavior is more generally cultivated as a mean to business owners' ends. A tricky point considering that the priority of businesses is to make money, not to make a better world. That is why it has become even more critical for consumers to assume their second key role in the path towards a more sustainable fashion. After having defined their inner values and what will make them stand as conscious and responsible consumers, they need to advocate their beliefs to transform their convictions into driving forces. If consumers keep buying just sustainable products that has been made responsibly, then the whole supply chain is going to change according to a diffusion model. And as a closed loop, consuming patterns will influence producers' actions, which will push all consumers' behaviors to change. Thus, consumers have the responsibility to send signals to make consumerism evolves while businesses have the duty to listen and answer these calls.

According to this logic, the industry's values will change to meet a system of responsible and demanding consumers. It implies that the way they do business and they communicate will become more trustful. In fact, we can observe that this movement has yet started to emerge, even if it still targets a niche. Many brands now understand the need to demystify Sustainable Fashion through a clear and easy access to information. Authentic transparency has become a real competitive advantage. To bounce few consumers' demand and engage a larger target, brands now delivered real solutions such as Everlane, this American company which based its entire business model on 'radical transparency' by shelling every retail price, explaining how much materials, hardware, labor, duties, transport cost as well as the margins made by the brand. Standards and labelling are also a way to educate consumers through information disclosure (Ethical Fashion Forum 2018). Today, it exists more than 100 different labels tackling environmental or social sustainability in the textile industry. The most recognized ones are the Oko-Tex standard looking at health standards and the European Eco-Label for Textile Products willing to reduce water pollution. Other labels address the issues of fair trade such as the FAIRTRADE Mark ensuring that 50% of the piece of clothing has been made with Fairtrade-certified materials. Of course, efforts remain to be done in order to clearly explain the meaning and differentiations between each label but it can strongly drive mass consumption to more sustainable patterns. Moreover, new labels need to be developed to cover the whole garment production process. An easier access to information has also been possible thanks to the launch of several apps that give simple rating system on clothing brands about ethical practices to help consumers in responsible purchasing decisions. Facing more attentive consumers towards traceability, companies need to prove authenticity in their communication campaign as well as creativity in sustainability policies. As the New York-based e-commerce brand, Zady, explained, «there is critical work to be done in creating not just a brand solution, but solutions that can turn the entire $2.5 trillion industry into a force for good». To transform the current linear economy into an efficient circular economy, companies need to show the example by considering the whole garment lifecycle (Vuletich 2016). Rather than focusing on offering ever more products, brands should guide consumers towards Slow Fashion by producing less but delivering more innovative services. The successful high-end outdoors company Patagonia has adopted this philosophy and launched in 2013 the Worn Wear program offering a service to repair any old clothes. This initiative has been promoted thanks to a «Worn Wear Tour» including a truck, the «Worn Wear Wagon», which went around the US to teach the benefits of putting on buttons or patches on old garments we wanted to throw away. Two workers were also present to offer a repairing service for more complex repairs. Patagonia's director of global communication, Adam Fetcher, explained that during this tour, they repaired more than thousands garments and they could teach people how important it is to make stuff lasts. The company, highly committed in its mission, also protests every year against the Black Friday operation, denouncing a buying binge phenomenon. For the 2016 Black Friday, they released 40 different repairs kits online while, the year before they organized many events to foster people to exchange their gear with others as a modern form of barter economy (Cheeseman 2016). And, this formula works pretty well: asking consumers to «Not buy this jacket» makes money. In 2015, Patagonia's turnover reached the amount of 750 million dollars, a growth of more than 10% in only one year (Vulser 2015). Regarding this success, many companies decided to adopt Patagonia's benefit corporation status and even joined the company's association, One Percent Planet. A project which aimed at educating companies towards positive commitments in social and environmental policies and refunding 1% of annual incomes to preserve earth. Today, this movement gathers more than 1300 companies. If Patagonia stands as a pioneer, many other companies actually advocate the need to re-frame our linear economy. The case study of Levi Strauss & Co. is a good example to show the high return on investment of a well implemented CSR communication strategy. Born in 1853, the company has gone through the years, knowing a glory golden era in the 90s. In 1996, the American family business' sales reached more than 6.2 billion euros. Nevertheless, in 2003, LS&Co's sales dropped to 3.6 billion euros (Kansara 2015). A decay mainly explained by the tough competitive climate led by Fast Fashion actors. The company reacted by implementing three core strategies: focusing on the timeless and iconic 501, promoting a true lifestyle and finally, standing against the core essence of Fast Fashion through a clear sustainable vision. A wise choice considering that the denim industry has one of the worst environmental and ethical footprints. In fact, it takes 1.7 million tons of chemicals to make 2 billion pairs of jeans per year. Levi Strauss & Co's Vice President of Sustainability, Michael Kobori, has demonstrated leadership by writing about the Life Cycle Assessment (LCA). This project aimed at analyzing the whole production cycle of a pair of Levi's 501 jeans to determine where they could improve their own behavior as well as consumers' ones. This research highlighted that much of the environmental damage actually occurs after the act of purchase. Therefore, Levi Strauss & Co. decided to provoke consumer change thanks to a well established CSR education programs. For example, the Care Tag for Our Planet Program's purpose is to teach consumers how to launder and care for their LS&Co clothing. They promote the «wash less, wash in cold, line dry and donate when no longer needed» moto (Kobori 2014). To create new products from old jeans, they initiated a collaboration with I:CO to put recycling bins in stores and collect clothing. Another collaboration with Goodwill, an American nonprofit association, calls consumers to recycle thanks to the «Give Back Box», a free shipping label to send back old clothes for online shoppers. Inspired by Patagonia's policies, Levi Strauss & Co. also launched the Levi's® Tailor Shop to follow the trend in fashion repair. This new service offers alterations, hemming, repairs and customs by experts. But the goal was not to put all responsibilities on consumers but also to act as a role model. Therefore, Levis also worked on its own production and distribution process. They launched the Water Less™ finishing techniques, which can reduce the use of water up to 96%, and a Worker Well-being initiative. Mentioning these two programs during an interview for Euromonitor International, Kobori stated: «We have seen reduced costs or improved business throughout our supply chain. We also know that younger consumers increasingly seek out companies that demonstrate social purpose and are more likely to buy from companies that support social and environmental causes». A fair assumption as Levis' sales have increased from $4 billion in 2003 to $4.9 billion in 2017. Thanks to its success, they can now invest in more R&D to find out solutions to recycle denim. They started a collaboration with Evrnu, a company which has developed a new technology to melt or dissolve recycled cotton to the cellulosic level to then create a new solid fiber. The criticism is unanimous, especially regarding LS&Co's Care Tag. The Levi's Facebook community as well as the Care Tag microsite received a warm welcome in 2010. In fact, the Care to Air blog post is Levi's 4th most visited post, with more than 3,100 views. The press also highlighted LS&Co's actions with 300 media stories and 500 million media impressions, including The NY Times, Fast Company and LA Times (Jack et al. 2011). Thus, the senior director of global media relations for Levi's, Kelley Benander, pointed out the huge benefits a brand can retrieve from being authentic with consumers. If Patagonia and Levi Strauss & Co. tackled the whole life cycle of producing and consuming clothes, others went with totally different ideas: stop buying, start renting. Rent the Runaway, Gwynnie Bee or even the French concept L'Habibliothèque have also experimented this new business model

(Drew and Yehounme 2017). For instance, Rent the Runaway, created in 2009 by Jennifer Hyman and Jennifer Carter Fleiss, is anchored in this new notion of «access economy», mainly fostered by companies such as Spotify and Netflix. From the early beginning, the company received $126 million in venture capital from several business angels such as Condé Nast and American Express. The immediately knew a huge success and in 2016, they managed to gather more than 6 million members, a work force of 975 employees and hit $100 M revenues.

All these innovative models and meaningful communication strategies to educate consumers would not have been possible without the internet. The means to engage new consumers is as crucial as the message to foster a collaborative movement and at a 2.0 era, what consumers need is to be charmed through viral and digital information. However, while «Millennials» is on every marketers' lip, the best strategy is in fact to look forward, towards the Gen Z—*born after 1995*, as by 2020, these digital natives will represent 40% of consumers. Thus, first and foremost consumers education should mainly uses mobile devices as 89% of them own a smartphone from their early 12 years old. Some companies have already seized this opportunity and launched mobile applications to educate consumers towards Sustainable Fashion. For instance, Done Good is a shopping app which filters brands to help consumers to make more responsible choices. Through the app, consumers can enter a product and select the important values they like, such as «supports workers», «green», «vegan», etc., then the app will display brands that offer what consumers are looking for. It has been created in 2015, and from now on this not so famous app has managed to divert more than $350k from big corporations' e-stores to DoneGood brands. In 2017, the startup even received the «Best for the World» awardee to promote the service and raise consumers' awareness. Moreover, the independent-minded Gen Z is convinced that its future is on its own hands and is therefore more willing to be part of a sharing economic model. Thus, committed brands should focus on being attractive and foster interactivity. To attract them, they should develop snackable and sharable contents, meaning creative visual contents spontaneously delivered through social media platforms such as Snapchat or Youtube, the two most favorite channels of Centennials. The philosophy of making a shift for yourself in our society is generally fostered by the greatest hope that others will see it and do it too. Sharing has even become a pathology. Fashion Revolution particularly understood how important it was to answer the need of Centennials for sharing and use this phenomenon to bring their call for more sustainability. In another format, Conscious Chatter also adopted this strategy by posting short podcasts. They talk about various issues such as fairtrade, Fashion activism or waste issues. These two concepts quickly communicate through emojis, gifs, short videos with bright colors and striking messages. Finally, this generation is also more seduced by humorous, relevant and authentic contents thus interactivity should be the key word to communicate on sustainability. Gamification, for example is a good way to engage consumers in becoming loyal to Slow fashion brands. If this strategy has not really been adopted by Fashion brands yet, the models of some green gamification apps can be a starting point. For instance, Recyclebank offers customers rewards for taking green actions such as recycling. These points can be redeemed with reward partners for home, clothing, gifts, etc.

Other ideas are to encourage green actions thanks to friendly competition or social comparison for instance.

Of course, these few examples of techniques to reach a larger clientele and deeply change the industry into a Slow Fashion model, would only work if consumers and producers show respect towards themselves, each others, the environment and the products. Then, slowly but surely, consuming and supplying behaviors will become more sustainable.

5 Concluding and Perspectives

To conclude, it appears that consumers, pure product of the second-half 20th century, considered as a genius invention created to revolutionize an out-of-breath economic and industrial model, has become a kind of modern day Frankenstein. The consumers' frenzy dance to systematically satisfy insatiable desires of goods, of more choice at lower price, gave birth to a chaotic whirlwind, so called the Fast Fashion industry.

Nothing could strike as a wake call except a drama, and that is what happened with the Rana Plaza collapse. A disruptive fact, a «disorienting dilemma» which initiated the first steps of a transformative learning process from brands, NGOs, governments and consumers. Nevertheless, if most of these passive actors tried to clean up their acts to feel less guilty, their willingness has never been enough authentic to start a concrete positive change. The «system», fascinated by its own «baby», wonders with blind eyes how to not radically change consuming behaviors that would make a total economy collapse, rather how to improve them to appear morally involved. The solution? Implementing an educating program made of moralizing discourses, new soft marketing strategies or responsible answers to surveys. Thus, Sustainable Fashion has always remained an utopian concept.

Offering new «green» products and feel-good massages have not done the job mainly because we forget that education is not a one-way process but a shared responsibility. Mezirow's theory is a starting point to think on how to turn passive consumers into critical and forward thinkers. The whole process is actually based on re-framing authentic values and widely communicate and collaborate to engage others. Thus, in a challenging environment, we will manage to tackle years of mass consumption and create a sharing circular system where communication and action will be adapted in order to raise united responsible actors.

The main hope for the future is that we would stop looking for making money off exploitive and vain consumer appetite. The solution for that is to learn how to spend our hardly earned money and how to generate profits from helping consumers and retailers in the real meaning of their lives. Buying less should lead the way to the creation of more innovative services and the production of less throwable products. We need to create a new kind of consumerism based on shared authentic values and higher needs (Booth 2016). Knowledgeable consumers as part of a transformative learning process can initiate this change by getting inspired of Emma Watson's words, known for her activism for women's right and more sustainability in Fashion: «If not me, who? If not now, when?».

References

Booth, M. (2016). *History: Consumerism*. https://www.youtube.com/watch?v=Y-Unq3R–M0&t= 316s. Accessed March 26, 2018.

Bowman, W. (2016). *Back in time: The history behind H&M*. http://theregister.co.nz/news/2016/1 1/back-time-history-behind-hm. Accessed March 13, 2018.

Cheeseman, G. M. (2016). *Educating consumer about buying sustainably*. https://www.triplepun dit.com/special/cotton-sustainability-c-and-a-foundation/increase-awareness-buying-sustainabl y-among-consumers/. Accessed April 7, 2018.

Chua, J. M. (2014). *Vanessa friedman: «Driving force of fashion is planned obsolescence»*. https://inhabitat.com/ecouterre/vanessa-friedman-driving-force-of-fashion-is-planned-o bsolescence/. Accessed 15 February 2018.

Coquery, N. (1998). Le march de luxe à Paris au XVIIIème siècle. https://clio-cr.clionautes.org/le-marche-de-luxe-a-paris-au-xviiieme-siecle.html. Accessed 24 February 2018.

Ditty, S. (2018). *Transparency is trending*. http://fashionrevolution.org/transparency-is-trending/. Accessed March 27, 2018.

Drew, D., & Yehounme, G. (2017). *The apparel industry's environmental impact in 6 graphics*. http://www.wri.org/blog/2017/07/apparel-industrys-environmental-impact-6-graphic s. Accessed March 24, 2018.

Ebeling, R. M. (2016). *The fable of the bees tells the story of society*. https://fee.org/articles/the-fa ble-of-the-bees-tells-the-story-of-society/. Accessed 24 February 2018.

Esthermsmth, (2017). *Transformative learning theory (Mezirow)*. https://www.learning-theories.c om/transformative-learning-theory-mezirow.html. Accessed April 1, 2018.

Ethical Fashion Forum. *Standards and labelling*. http://www.ethicalfashionforum.com/the-issues/ standards-labelling. Accessed April 5, 2018.

Friedman, V. (2013). *How the IWWIWWIWI generation is changing fashion*. https://www.ft.com/c ontent/fc6d164f-0afb-3424-ae15-2f707cedc52d. Accessed March 26, 2018.

Goodwin, N. (2008). *Consumption and the consumer society*. http://www.ase.tufts.edu/gdae/edu cation_materials/modules/Consumption_and_the_Consumer_Society.pdf. Accessed March 12, 2018.

Hess S. (2011). TEDxSF—Millennials: Who they are & why we hate them. https://www.youtube. com/watch?v=P-enHH-r_FM. Accessed 27 March 2018.

Howie, P., & Bagnall, R. (2013). A beautiful metaphor: Transformative learning theory. *International Journal of Lifelong Education, 32*(6), 816–836.

Inditex. (2017). *Annual Report 2016*. https://www.inditex.com/documents/10279/319575/Inditex+ Annual+Report+2016/6f8a6f55-ed5b-41f4-b043-6c104a305035. Accessed March 27, 2018.

Jack, D., Benander, K., & Nash, J. (2011). *Case study: Levi's sustainability initiative proves that a solid CSR message does wash with the public*. http://www.prnewsonline.com/case-study-levis-sustainability-initiative-proves-that-a-solid-csr-message-does-wash-with-the-public/. Accessed April 8, 2018.

Kansara, V. A. (2015). *Levi's se remet en selle*. http://www.lemonde.fr/m-styles/article/2015/02/2 4/levi-s-se-remet-en-selle_4581768_4497319.html#meter_toaster. Accessed April 8, 2018.

Kobori, M. (2014). *Why consumers education is vital for corporate sustainability success*. http://www.sustainablebrands.com/news_and_views/behavior_change/michael_kobori/wh y_consumer_education_vital_corporate_sustainability_. Accessed April 8, 2018.

LeBlanc, R. (2017). *Textile recycling facts and figures*. https://www.thebalance.com/textile-recycli ng-facts-and-figures-2878122. Accessed March 25, 2018.

Martin, S. (2017). *Qu'est-ce due le fast fashion?* https://www.solenemartin.com/13-quest-ce-que-l e-fast-fashion/. Accessed March 13, 2018.

Monneyron, F. (2006). *La sociologie de la mode.* Que Sais-je?, Paris.

Parry, S. (2016). *The true cost of your cheap clothes: Slave wages for Bangladesh factory workers.* http://www.scmp.com/magazines/post-magazine/article/1970431/true-cost-your-cheap-clot hes-slave-wages-bangladesh-factory. Accessed March 24, 2018.

Perry, P. (2018). *The environmental costs of fast fashion.* https://www.independent.co.uk/life-style/f ashion/environment-costs-fast-fashion-pollution-waste-sustainability-a8139386.html. Accessed March 25, 2018.

Pringle, B. (2018). *Millennials demand socially responsible clothing, but won't buy it.* https://www.washingtonexaminer.com/millennials-demand-socially-responsible-clothing-bu t-wont-buy-it. Accessed March 27, 2018.

Rahmil, D. J. (2014). *What is a «client» of the sharing or collaborative economy?* https://digital-society-forum.orange.com/en/les-forums/368-quest-ce_quun_l_client_r_d e_leconomie_collaborative. Accessed April 3, 2018.

Saner, E. (2017). Sustainable style: will Gen Z help the fashion industry clean up its act? https://www.theguardian.com/fashion/2017/apr/25/sustainable-clothing-fashion-revolut ion-week-rana-plaza-emma-watson. Accessed 27 March 2018.

Vince, G. (2012). *The high cost of our throwaway culture.* http://www.bbc.com/future/story/20121 129-the-cost-of-our-throwaway-culture. Accessed February 13, 2018.

Vuletich, C. (2016) *How to engage with ethical fashion.* TEDxSydney. https://www.youtube.com/ watch?v=WXOd4qh3JKk

Vulser, N. (2015). *La bonne recette de Patagonia, société (presque) verte.* http://www.lemonde.fr/ economie/article/2015/12/10/la-bonne-recette-de-patagonia-societe-presque-verte_4828877_32 34.html. Accessed April 7, 2018.

Westerman, A. (2017). *4 Years after the Rana Plaza tragedy, what's changed for Bangladeshi garment workers?* https://www.npr.org/sections/parallels/2017/04/30/525858799/4-years-after-r ana-plaza-tragedy-whats-changed-for-bangladeshi-garment-workers. Accessed March 24, 2018.

Printed in the United States
By Bookmasters